The Accidental Engir

A life leading to the design of the World's First Microprocessor created for the U.S. Navy F-14 "Tomcat" supersonic fighter jet

The Accidental Engineer

A life leading to the design of the World's First Microprocessor created for the U.S. Navy F-14 "Tomcat" supersonic fighter jet

Autobiography of Ray M Holt

The cover picture

The cover picture is the team that started the microprocessor customer training for Intel, circa 1973. From the left: Manny Lemas, President and Co-Founder of Microcomputer Associates; Bob Garrow, Intel marketing; Phil Tai, Intel Marketing; Ray Holt, Executive Vice-President and Co-Founder Microcomputer Associates Inc.

Intel realized that engineers did not know how to program and so contracted Manny and I for two years to travel the USA teaching Intel microprocessors. The last day of the 5-day course was with Gary Kildall on PL/M, his high-level Programming Language for Microcomputers (PL/M).

Bob Garrow left Intel soon after this and started Convergent Technologies which later was bought by Unisys in 1988. In 1974 Gary Kildall started Digital Research and created CP/M, MP/M, and PL/M, the first popular operating systems for personal computers.

Checkout **FirstMicroprocessor.com** for more stories, books, photos, and details.

Version 2.00 – August 25, 2022 (US Version)
ISBN 978-1-4710-7895-8

©2006-22 Copyright by Ray Holt All Rights Reserved   2nd Edition
This publication cannot be reproduced in any form, without the written permission of Ray Holt.

INDEX

The cover picture

Author's Preface
Tribute to the F-14 CADC team
Recollections from Jim Kawakami and Ken Rose
Legacy
Acknowledgements
Book Preface
Foreword – Darold Cummings, Top USA Aerospace Designer
Foreword – Bart Everett, Commander of Military Robots
Foreword - Dave "Bio" Baranek - Former F-14 RIO and Top Gun instructor

Chapter 1. Baseball, Radios, and Engineering
Chapter 2. The Accidental Engineer
Chapter 3. The Navy Calls: The F-14 Brain (with text by Tom Redfern)
Chapter 4. Technology vs Forty Square Inches
Chapter 5. The Danger Zones
Chapter 6. My Claims
Chapter 7. Beyond the F-14: More Microprocessor
Chapter 8. Microprocessor / Microcomputer Projects
Chapter 9. JOLT and Super JOLT: Birth of a Home Computer
Chapter 10. VIM-1, SYM-1, & SYM-2
Chapter 11. The Christian Athletic Association Inc (CAA) ... and other ventures
Chapter 12. What is Ray doing now in 2022?
Chapter 13. Vision: STEM and Robotics in the 21st Century

©2006-22 Copyright by Ray Holt All Rights Reserved   2nd Edition
This publication cannot be reproduced in any form, without the written permission of Ray Holt.

Author's Preface

Growing up was not so normal. Our father was an Itinerant welder and pipe fitter and we traveled across Kansas, Oklahoma, Texas, Colorado, Arizona, New Mexico and California. We had a family of five in a 20-foot trailer and moved about 15 times during our elementary school age. Waking up in a different city and school two or three times a year was not unusual. Coincidently, we were in Compton California when each of us was born. Finally, when I was 10, we stayed in Compton and two years later our parents bought their first house for $5,000. This was quite exciting for all of us. We attended Stephen Foster Elementary School, Whaley Junior High, and Dominguez High School. Yes, we walked to all the schools most of the days. Dominguez was four miles each way.

During this very portable time of our life, it did not seem to affect my sister, Liz, who was two years older or my brother, William "Bill", who was 1-1/2 years younger. Both of them were excellent, straight-A students and excelled in extra curriculum activities, and earn many top awards in high school. In junior high Liz won an art contest and a trip to Washington DC. Bill won a sports and academic scholarship to Stanford University and was selected for Boy's State. I seemed to come through this portable time with fewer skills. I struggled with reading, comprehension, and grammar. I took English grammar twice and was asked to leave Spanish. I was able to maintain an overall "B" average and did well in Math. I had no direction leaving high school other than I was told "do not go into engineering". I did the only natural next step and that was to attend the local Compton Community College. That turned out to be quite a challenge.

©2006-22 Copyright by Ray Holt All Rights Reserved   2nd Edition
This publication cannot be reproduced in any form, without the written permission of Ray Holt.

There are times in one's life where you can't see to the end of the day much less what you will do in life. During my teens and early 20's I had lots of those days. Just to mention some of them to make my point; walking out of college and not doing any paperwork to drop classes (I received all F's) eventually made me appreciate college administration, standing on dirty, messy, smelly garbage and handling a fire hose all day makes you wonder what just happened to your life and makes college look pretty good. Moving very far away to a university with no real goals or plans puts a huge void in front and back of you. Placing 2nd in an axe throwing contest, when you had to be shown how to hold the axe, was not supposed to happen. Having been told by your college major dean to take a class outside your major and having it change your college direction and location. Taken a course you had no idea of the application of the subject and having it become the foundation of your career. Taking six years to finish college and wondering what you just learned. Walking into a job thinking you were going to be doing one thing and all of sudden you are told you are doing something else that is beyond your thinking.

All of us are faced with opportunities, either from bad decisions on our part, from opportunities just because we were there, or from opportunities because someone decided it made sense for us. All through these opportunities we are just going for the ride and not even trying to fit it all into a plan as we really don't have a plan. Sometimes life takes us along a very strange path that makes no sense to us. For those that have a spiritual background and have learned to depend on God or had committed to God's direction earlier in life then we just have to know that He created these opportunities for our good. I often wonder why He doesn't just ask us, however, if He did, we would probable just laugh and rebel, which is the very reason He guides us in the first place.

©2006-22 Copyright by Ray Holt All Rights Reserved 2nd Edition
This publication cannot be reproduced in any form, without the written permission of Ray Holt.

I was the most unlikely person at 24 years of age to be a key designer on a highly state-of-the-art computer system for a very important state-of-the-art fighter jet. Not just any fighter jet but one that would become critical in defending our country and would be exposed as a top-notch fighter jet in the eyes of many people. Here are a few reason why I was the most unlikely to be one of the designers; 1) I had never designed a computer before, 2) I had never worked for the military engineering establishment before, 3) I had only taken ONE class that might be associated with computer design, 4) I was only 24 and still would rather play baseball, 5) My college friends that helped me get through college were not here, 6) I was told I should not talk to others outside the company about what I am about to do, 7) I was completely lost in all the military acronyms and had no idea which applied to me or not. And then there was the high school counseling words "Do not go into engineering." and the discouragement of knowing my mechanical aptitude is low. And now, all of a sudden, I am to engineer design a state-of-the-art digital computer (not done by anyone else) that does the same thing as a previous mechanical computer. So, in summary, I did not know where I was to go and I had no idea where I came from.

This is how it all started. "Lord, you brought me here on this long path of opportunities so now you are going to have to help me finish this project.... So, let's go." For those of you reading this that might feel like they are living a little part of this story then I would say just increase your faith and just trust in the Lord as he knows what He is doing. I would also say there is no reason to look back if you are surrounded by people that are depending on you. That's what the Garrett AiResearch team did for me.

Here are some wonderful outcomes from this project. When I was hired the company did not have the contract. I was told to take two months and learn all I can; read, read, practice designs, study, and attend classes. I was sent to a premier computer arithmetic class at UCLA with a very well-known instructor named Professor Gerald Estrin. Professor Estrin retired in 1991 Emeritus as the Department Head of the UCLA Computer Engineering and Computer Science.

I was given all the equipment and electronic components I needed. I was assigned the best Electronics Lab Technician, Lynn Hawkins, who bailed me out on many occasions. I was given freedom to hire programming help with included James Lallas and my brother Bill Holt. I was given freedom to choose the final technology for the computer (and thus the responsibility that it really would work). I was given access to the world's best applied mathematician, Bill McCormack, who could analyze complex equations in his head and then give you the adjusted "constant" to make everything work fine (still amazing today). I was surrounded by the best excellent experienced aerospace engineers I could have ever imagined that helped me make the computer system actually work in the required space and in the required environment. This team was led by Tom Redfern and Ralph Ichikawa with designers Cameron Pedego, Russ Almand, and Dave Knickerbocker. My logic design partner was Steve Geller. Steve was a high-level system designer and did not like details or hardware. He received a lot of jokes about that. However, Steve's experience and contribution of the big picture was very invaluable and highly contributed to the project's success. And finally, I was teamed up with world-class integrated circuit chip engineers at American Microsystems; Ken Rose, Jim Kawakami, Brian Schubert, and Gordy Leighton. Ian Linton, a Test Engineer at Garrett, worked closely with me in making sure all the final testing really proved the computer was doing the job it was supposed to do.

Yes, I had many days that seems like the computer was impossible to make. Too much math to perform, too little space for the microchips, too new a technology to know if the microchips would actually work (of course on paper we proved it would work). Being told, at least twice, the microchips are too large to make and forcing us to re-architect the system design. Too little time in a day to meet deadlines. And finally, who was I to say this was all going to work or not. There was no history on how to prove the math would work, the computer logic was correct, the new technology chip design and layout would work, the state-of-the-art 8-layer printed circuit boards were a risk, and I was the least likely person to be assigned to this computer design project.

Many days I felt God literally controlling my brain. This is especially in the area of the detail computer design. This stuff gets really involved and complicated and when you get tired it all gets very fuzzy. Making organized sense out of 1000's of logic gates being moved by electricity over 300,000 times a second gets challenging and difficult. Every logic gate had to work correctly, at the right time, in order for the computer to be considered ready to go. Sometime just pausing and taking a deep breath gave me a jolt of temporary energy and sanity and a new direction or a new design technique. Sometimes in the shower some really cool solutions to difficult design techniques came to me. Sometimes opening some of my recently acquired books (books by Gordon Bell and Yahoan Chu) highlighted a solution to difficult design areas. These microchips used several unique design techniques that were the result of the above God moments. These design techniques I mention below.

Besides the challenges, there were many success moments. 1) The first I remember was when I was successfully able to simulate the complex logic of the 20-bit binary divider and binary multiplier. Exhaustive test patterns were run against the simulated logic design. 2) When the prototype of the microchips worked. 3) When I was told that American Micro Systems would make the microchips. Initially, all vendors declined to make the chips. 4) When the first microchips worked and worked and worked. What a joyous day after two years of hard, tedious work. All the chip logic worked perfectly the first time. 5) And finally, the day the first CADC was delivered to Grumman Aircraft.

My whole story goes from being lost, following blindly (by faith), accepting opportunities, moving forward with what I had, working with others, seeking God's direction, and finally letting Him work out the complex details. In the end it was my stubbornness to not accept failure that drove me to pursue excellence and, of course, doing what an engineer does and that is to make sure every detail of their design works under all required conditions. In a military environment those conditions included wide temperature extremes, vibration, shock, and acceleration, and, of course, design integrity.

I get asked a lot what it was like keeping the F-14 CADC a secret for 30 years (until 1998) while yet still working as a designer and developer in the pioneering microprocessor and microcomputer industry. Initially, not being able to present my first paper to Computer Design magazine, was very frustrating as it would have helped my career. Not being able to patent the many unique and first techniques was difficult to understand. That would have helped my future finances. I have received no royalty from the microprocessor nor have I taken any kind of stipend for speaking about it since. Especially, since years later others were able to obtain the patents and royalties. The military insisted no patents as at this time all patents were public and the project was secret. Today that is not the case, patents can be marked as private.

©2006-22 Copyright by Ray Holt All Rights Reserved   2nd Edition
This publication cannot be reproduced in any form, without the written permission of Ray Holt.

After the first two years it was not as difficult as other microprocessors had been announced. I had completed two others with American Micro Systems (the AMI7200 and 7300). In 1972 with the announcement of the Intel 4004 and then 8008 and 8080 and with the big marketing push it became very hard to keep quiet. Since I was a consultant to Intel it was hard to not mention to them that they were not first, not even close. Once I got over that initial frustration the rest of my time at Intel became easier. I also realized that I had a lot of knowledge others did not have and that I should just make the best of the many opportunities. During the 80's and 90's I did not think much about it or talk much about it except to my family. I think the overall disappointment did affect my personality and probably made me a little difficult to be around at times.

I still see myself as the most *unlikely* person to have designed the F-14 CADC microchips, however, God knew I could do it and with His help it was a highly successful, reliable, and dependable flight computer for the highly successful F-14 Tomcat. He developed a personality of "perfection" in me that insisted that every detail worked. I had to be convinced on paper and in hardware and software simulations that every one of the 10,000's of transistors and logic gates worked perfectly and under all of the environmental conditions.

My hats off and much praise to all the F-14 pilots and RIO's that risked their lives in this huge machine guided by my design. I have yet to read about a CADC in-flight failure. Job well done is nice to hear but no CADC in-flight failure is music for a lifetime.

To the entire design team from Grumman, Garrett AiResearch, American Microsystems, and all the other vendors of the F-14... job well done and may the music keep playing.

©2006-22 Copyright by Ray Holt All Rights Reserved   2nd Edition
This publication cannot be reproduced in any form, without the written permission of Ray Holt.

F-14A CADC Hardware Simulator

Tribute to the F-14 CADC team

The real legacy of the F-14 CADC was in all the future careers of the design engineers.

The Garrett AiResearch and American Microsystem, Inc. (AMI) design teams.

Garrett AiResearch

Paul Lyons – Department Manager

Andy Papadeas – Director of Engineering

Bill McCormick – Mathematician/Analyst

Phil Erath – Project Director

Dick Barcus – CADC Manager

Steve Geller – Logic Design Engineer

Ray Holt – Logic Design Engineer

Larry Hammond – Project Documentation Manager

Cleve Hildebrand – Circuit Design Director

Ralph Ichikawa – Circuit Design Manager

Tom Redfern – Circuit Design Manager

Russ Almand – Circuit Design Engineer

Cameron Pedego – Circuit Design Engineer

Dave Knickerbocker – Circuit Design Engineer

Ian Linton – Test Engineer

©2006-22 Copyright by Ray Holt All Rights Reserved   2nd Edition
This publication cannot be reproduced in any form, without the written permission of Ray Holt.

Lynn Hawkins – Engineering Technician

K.T. Chang – Programming Manager

Bill Holt – Programmer, Diagnostics

Jim Lallas – Programmer, Simulation

Murray Lubliner – Programmer, Simulation

Jessica Kuo – Programmer

C.Y. Chin – Programmer

Pete Miller – Test Programmer

Al Gaede – Test Programmer

American Micro Systems, Inc. (AMI)

Ken Rose – Director of Engineering

Al Pound – Design Manager

Jim Kawakami – Project Engineering Manager

Brian Shubert – MOS Design Engineer

Gordy Leighton – MOS Design Engineer

Lloyd "Red" Taylor – Research Engineer

Jay Miner – Research Engineer

Recollections from Jim Kawakami
AMI Project Engineering Manager

"Most of the commercial designs of CADC era were intended for calculators. My recollections are a little fuzzy but when I started working in 1965 at General Micro Electronics (a Philco Ford company) the most complex design was a calculator for I think Smith Corona. At American Micro Systems, the LSI type designs were all for either calculators (Ricoh, Burroughs) or government projects for NSA, Garrett, etc. I remember second generation Ricoh and Burroughs (France) calculator chip sets both utilizing computer fundamentals, i.e., buses, instruction sets, programmable memory, ALU, etc. The Intel 4004 was a custom design for Busicom in Japan and I think the lead designer for the 8080 came from Busicom. It is too bad that Intel was able to capture all the first microprocessor hype with their 4004. In the commercial microprocessor market, they were first and won the day. I recall that the second-generation Motorola microprocessor (6800) was deemed superior to the 8080 but Microsoft picked the 8080." **Jim Kawakami**

Recollections from Ken Rose, AMI Director of Engineering

"Early in 1966 GMe was sold to Philco-Ford and a number of engineers left in late spring to form a new company, AMI, American Microsystems Inc. under Howard Bobb, President. I personally decided to leave the government and join AMI in August, 1966. The VINSON program was subsequently stopped at GMe and given to Texas Instrument for completion.

1967 - AMI, under NSA sponsorship, developed a fairly detailed course and textbook in the techniques of designing MOS-LSI circuits, including system design considerations, logic, circuit and layout techniques as well as the rudiments of circuit fabrication. The course was taught to a number of people, government and contractors, establishing a fairly large group capable of designing custom LSI circuits for both government, NSA Comsec, NASA and various weapons systems, and commercial use, primarily electronic calculators. At least some of the companies that had students in this course were RCA, Burroughs, National Cash Register (NCR), Honeywell, TRW, and Collins Radio.

1967-71 - A number of programs initiated by the agency were intended to take advantage of the new LSI technology and to do this, AMI subcontracted for a number of the companies involved with developing the equipment in order to do the MOS-LSI circuits. I led the teams at AMI that did the circuit development for the INY program (5 circuits), a space Comsec equipment developed by TRW, the PARKHILL speech scrambler (8 circuits) developed by Collins Radio, Newport Beach, Calif, and the FOXHALL program (13 circuits) developed by Honeywell, St Petersburg, Fla, the predecessor to the Army/Air Force TRITAC program. These programs were all based on custom circuit designs, with standard MOS-LSI products not appearing until the early or mid-70's. Memories were the first to be developed as standard items, with microprocessors to come next, being derived from electronic calculator designs." **Ken Rose**

Legacy

Ken Rose (AMI) continued his work in the cryptographic field for the Department of Defense. Tom Redfern (Garrett AiResearch) continued his work with National Semiconductor in CMOS design and was selected as the first National Fellows for design excellence. Jim Kawakami (AMI) continued his work with AMD and became a top innovator and leader in microprocessor design. Brian Schubert (AMI) continued his work with AMD and Intel in graphics chip design and led the Intel Graphics Division.

Jay Miner (AMI) left AMI in 1970 and joined Altair and became the father of the Amiga. I would say that is probably the single most important contribution of most designers in the entire industry.

Ray Holt (see Ray's Resume at www.FirstMicroprocessor.com) not only designed several other microprocessors, co-founded Microcomputer Associates, Inc. and co-published the industry's first magazine, The Microcomputer Digest, he was a contributor at most technical trade shows. He also was hired by Intel in 1974, along with Mr Manny Lemas, to travel the USA teaching engineers how to program the 4004, 8008, 8080 and PL/M language. The big reason for this was that, and I quote from Intel Marketing, "We are having a difficult time selling the products because engineers are not familiar with programming their design." It might be safe to say the Mr Ray Holt and Mr Manny Lemas were big contributors to the success of the sales of the Intel 4004, 8008, 8080 and PL/M. Mr Gary Kildall, designer of PL/M and later CP/M was also a course instructor.

Gary was also a huge contributor to the courses as he would come in the last day or two and teach PL/M which most engineers could understand because it was a high-level computer language and contained words that were somewhat normal. In my opinion Gary Kildall was one of the greatest technical and business contributors to the small computer industry and market. Jay Minor (mentioned above), Chuck Peddle and John Fagans from Commodore were the other great contributors.

Acknowledgements

There are many people that influenced me during my life and career. Needless to say, my family (Mom, Dad, Bill, Liz), my wife (Lynda), and sons (Mark, Michael, Brett). No one discouraged me during the development of the F-14 CADC. Certainly no one understood the work that was being done, even if I could have stopped and explained it in detail. The technology was so new and the application of the technology kept the project moving like a ball. They also had to put up with me keeping this a secret for 30 years while the world took the largest technological advance ever and I could not talk about this until 1998.

Let's start with the entire Central Air Data Computer (CADC) team. This was a team effort and the microprocessor was only a part of the entire CADC project. This was the most powerful set of engineers and programmers I have seen together in my entire career. Thank you to all of them from Garrett AiResearch and American Micro System. Their own lives and careers proved their excellence.

To Chanda Roby, Mississippi, now President of REAL Christian Foundation, who put in hours and hours encouraging and working with me to get the manuscript started and to help me to put it in a readable form. She was my big motivator to get moving on this. Thank you, Chanda.

To Amy Skalicky, Colorado, who put in hours making proposals to agents and publishers trying to get one of them to see that this is an important story to tell. Thank you, Amy.

©2006-22 Copyright by Ray Holt All Rights Reserved   2nd Edition
This publication cannot be reproduced in any form, without the written permission of Ray Holt.

To Leo Sorge. The one who understood the importance of this story in technology and to history and ran with it as Chapter One in his own book, "From Dust to the Nano Age", and finally his great skills and effort in putting together what you now have in your hand. Thank you, Leo.

Finally, to all the brave F-14 Tomcat pilots and RIO's that trusted this high technology computer to take them on 1000's of high-risk missions and back without a hitch. Your stories are the most important. I salute all of you!!!

©2006-22 Copyright by Ray Holt All Rights Reserved   2nd Edition
This publication cannot be reproduced in any form, without the written permission of Ray Holt.

Book Preface

This is the story as I know it of my life and adventures leading up to and designing the Central Air Data Computer (CADC) for the F-14 Tomcat airplane. The CADC was the main computer that controlled the moving surfaces of the airplane and provide the critical data to the pilots and the weapons systems.

The F-14 turned out to be one of the most successful fighter jets in American aircraft history having performed its duties for over 35 years, the longest life of any American military aircraft.

I am starting this story when I was young because many events or failures in my life lead me to the F-14 project. Every one of them is important to the direction and decisions I made. When I started the F-14 computer design I was 24 years old and had just graduated from college with a bachelor's degree in Electronics Engineering. I did not know what I was about to be asked to do.

The project took two years to complete, 1968-1970, and upon completion of the F-14 computer design, I attempted to document the effort in a technical paper for Computer Design magazine, the premium magazine for computer designers. The paper was accepted; however, I could not get approval from the US Navy to allow it to be published. After several attempts over the next many years, I was finally able to get approval in 1998 (thirty years later).

The major significance of the F-14 CADC was that the design of the computational chip set is, arguably, the first working microprocessor set chip. Many technical and technological feats were accomplished. Whether this is of technical or historical importance is for history to deal with.

©2006-22 Copyright by Ray Holt All Rights Reserved   2nd Edition
This publication cannot be reproduced in any form, without the written permission of Ray Holt.

The first public announcement of the F-14A CADC was a published article by the Wall Street Journal on September 22, 1998. This paper and the details of the design were first presented publicly by myself at the Vintage Computer Festival held at the Santa Clara Convention Center on September 26-27, 1998.

Many thanks to Mr. Sam Ismail of the Vintage Computer Festival for, not only his believability of this design, but for his hard work in making it a significant announcement in the microprocessor world.

Needless to say, a major announcement 30 years late, causes many good people to say and do many strange things and for many academic professors and design purist to doubt if what really happened did happen, or is it "just a good story told too late" as some say. Some argue that the marketing of a product makes it great or how many web searches it creates, or even how long it's promoted on page one of google. Certainly, marketing is important but technical excellence never takes second. Is accurate history ever too late?

My purpose is to document how it all happened, as for me, it is not too late and it is an important and good story of great importance.

Foreword by Darold Cummings.
Top USA Aircraft Designer, 2015

There is a very interesting book entitled "The Nearly Men" by Mike Green. It chronicles the inventors who should be given credit for major inventions, but failed to gain recognition for various reasons, including of lack of a credible paper trail, lack of funds to defend their intellectual property, or outright deception. Most of these errors in attribution would not be possible today due to the ubiquitous environment of instant electronic information. Ray Holt's story about the invention of the microprocessor has a similar Machiavellian flavor, in that the key information that would have established Ray's claim as the developer of the first practical microprocessor was withheld from the public for 30 years because of military security reasons.

Ray Holt's story about the invention of the microprocessor is really only a segment of a much larger story arc that could easily have started as a movie called "Straight Outta Compton", if it had been filmed in the late 1950's and early 1960's. My connection with Ray began at Dominguez High School in Compton, California, in the late 1950's. At that time Compton was a very blue-collar, middle-class community. Most families had a working father, a stay-at-home mother, and several siblings. My family's house in Compton cost $4000, and was very small by today's standards, but we really did not know any better, and seemed very fine to us. I got along very well with my parents, and our house was a popular hangout for the neighborhood kids. Large American-made cars provided transportation, vacation freedom, and a visible sign of success. A two-car family was very unusual. Walking or riding a bicycle was the normal mode of transportation for most students.

My father preached a very strong work ethic, and when I was about 14 he wanted me to find some type of work to do on the weekends. He bought me a gas-powered lawn edger (which was very uncommon at the time), and I edged

©2006-22 Copyright by Ray Holt All Rights Reserved   2nd Edition
This publication cannot be reproduced in any form, without the written permission of Ray Holt.

lawns on the weekends. I charged fifty cents to edge a lawn, and was usually able to make five dollars on a weekend. I was able to pay my dad back the $30 he paid for the edger within a few months, so after that I was in the pure profit stage. Unfortunately, I had hay fever, and after a long day of edging I was dirty, sneezing, and my eyes watered quite badly. Like Ray's job of watering garbage, I knew that this is not what I wanted to be doing later in life!

I met Ray when we were on the Dominguez High School wrestling team together. Most of our classmates could not wait to graduate and go straight into working as an automobile mechanic, barber, construction worker, plumber, welder, or assembly line worker at an aircraft plant. North American Aviation, Douglas, Hughes, and Northrop had nearby plants that were in full "Cold War" production operation. Both Ray and I had plans to go to college, which was a minority view in Compton at the time. There was one particular class at Dominguez that had a big impact on my career. In my senior year Dominguez offered a drafting class, taught by a Hughes Aircraft employee. The summer after graduation, I was able to get a job at North American Aviation as a draftsman because I had taken this class. This was my introduction to a 50-year career in Aerospace!

My grades were good enough for a State College, but for financial reasons, it was only feasible to live at home and commute to school. Long Beach State College (now know at California State University, Long Beach) was the closest campus, and was quite beautiful, with plenty of green open space (most of which is now covered with buildings!). I really wanted to be a car designer, so I signed up for the Industrial Design degree. Ray's college journey turned out to be much more desultory than mine. Ray and I both graduated from Dominguez in 1962, and the epithet of "Compton" changed soon after. The Watts Riots in the summer of 1965 spilled over into Compton, and the very nature of the city was rapidly changed. I remember telling people in college I was from Compton, and they asked what gang I was in!

©2006-22 Copyright by Ray Holt All Rights Reserved   2nd Edition
This publication cannot be reproduced in any form, without the written permission of Ray Holt.

Ray went from a disappointing stint at the local Compton Junior College to the University of Idaho (U of I) in Moscow, Idaho. As an interesting part of the story arc, I now live in Coeur d'Alene, Idaho, and U of I is considered a local university. Ray's intent was to study Forestry (a great thing to do in north Idaho!). A class in the physics of electricity changed his personal compass heading yet again! Ray decided that working with electrons was really in his future, and moved back to California and attended Cal Poly (California Polytechnic University) in Pomona. When Ray graduated in 1968, he took a job at Garrett AiResearch in California.

After college graduation in 1967, I accepted a position in Advanced Design at North American Aviation in El Segundo, California. One of the first projects I worked on was the FX, which was a new fighter design for the U.S. Air Force. At the same time, Ray was working at Garrett AiResearch on the microprocessor for Grumman entry for the U.S. Navy VFX program. In July 1968, the Naval Air Systems Command (NAVAIR) issued a request for proposals (RFP) for the Naval Fighter Experimental (VFX) program. VFX called for a tandem-seat, twin-engined air-to-air fighter with a maximum speed of Mach 2.2. In the tandem seat arrangement, the pilot was in the front seat, and the RIO (Radar Intercept Officer) was in the back seat. The RIO tracked targets on the radar, and as many as 24 targets could be tracked simultaneously, and six could actually be engaged with air-to-air missiles at one time.

Grumman was selected as the winner of the VFX program in January of 1969, and the aircraft was designated the F-14, which became the star of the movie "Top Gun". The F-14 wing sweep could be varied between 20° and 68° in flight, and was automatically controlled by the Central Air Data Computer (CADC, Ray's microprocessor), which maintained the wing sweep at the optimum lift-to-drag ratio as the Mach number and angle-of-attack changed. The CADC also controlled the slats (control surfaces on the front of the wing) and flaps (control surfaces on the back of the wing) for take-off, landing, and during combat maneuvering. This was a huge advantage for the Grumman design, as the pilot could not do this on his own, and was a key to the win. The F-14 went through several upgrades to the engines and avionics, and even mission priorities. The F-14 had its first flight in 1970, and was finally retired by the Navy in 2006, having a long and successful career. This was the first chapter in Ray's quest to be recognized as the inventor of the microprocessor. When the F-14 CADC project was declassified in 1998, Ray was finally able to stake his claim to being the inventor, and even made it into the Wall Street Journal.

Fast-forward a half century to complete the story arc. Dominguez High School was hosting the 50-year anniversary reunion for the class of 1962. I had missed the class reunions for the last 20 years, mostly because I race land speed records at the Bonneville Salt Flats every August at the same time the reunions occurred.

I received the usual flyer about the reunion, and decided to contact 1962 classmate Barbara Beacon about information on Ray, as he was the one person, I really would have enjoyed seeing at the 50-year Reunion. Barbara made the connection, and Ray contacted me and described his new life in rural Mississippi helping low-income children learn about engineering, math, and robotics. Ray yet again in a whole new setting! I had retired from Boeing as a Technical Fellow eight years earlier, moved to north Idaho, and was

Darold at Bonneville Salt Flats 2012

consulting for various aerospace companies and government agencies, and also presenting Creativity Workshops to universities around the United States. So here we were, both approaching 75 years of age, but actively pursuing activities in very technical areas. Obviously "retirement" did not appeal to either one of us!

Ray and I connected a few years later in Dallas, Texas. I was notified by the American Institute of Aeronautics and Astronautics (AIAA) that I was being awarded the 2015 AIAA Aircraft Design award certificate and medal at the AIAA Aviation Conference in Dallas, Texas, in June of that year. The inscription is shown below:

> American Institute of Aeronautics and Astronautics
> Aircraft Design Award 2015
> **Darold B. Cummings**
> For a career demonstrating exceptional skill and creativity in the configuration and design of aircraft, and inspiring future generations of aircraft designers.

This was a very high honor for me, and Ray told me he would like to come and see the presentation. Ray came to the ceremony, and was gracious enough to post a You Tube video of the presentation. At this point, it was the first time I had seen Ray in 53 years! Ray showed me all his documentation on the development of the F-14 microprocessor, which was extremely fascinating, since I was in the aircraft design business when the F-14 was introduced. I gave him an overview of the projects I was working on, and the Creativity Workshops that I present to universities. Ray wanted to know if I could come to Mississippi and present my Creativity Workshop to a variety of groups, including high school students, college students, and teachers. I was delighted to again be part of Ray's life, and I made my first presentation in Mississippi in 2017. The Workshop was very successful, and I was truly inspired by the enthusiasm Ray had instilled in both the students and the teachers. The 55-year story arc was complete!

Foreword – Bart Everett – Commander of Military Robots

I first met Ray Holt in 2004, when he came to visit the Unmanned Systems Group at what is now the Space and Naval Warfare Systems Center in San Diego, CA. Ray had stumbled across one of our many technical articles on the *ROBART* series of autonomous security robots. Back in the early '80s, I had used the versatile Synertek *SYM-1* single-board computer in my first two prototypes of this series, *ROBART I* and *ROBART II*. As Vice President for Engineering at Synertek Systems in Santa Clara, CA, Ray had been the hardware designer for the *SYM-1* in 1978, hence his interest. As I was soon to learn, he had also designed the fly-by-wire flight computer for the Navy's much accomplished *F-14 Tomcat*, featured in the popular movie *"Top Gun"*, starring Tom Cruise.

Robart I, II, and III

http://www.theoldrobots.com/OddsEnds.html

©2006-22 Copyright by Ray Holt All Rights Reserved   2nd Edition
This publication cannot be reproduced in any form, without the written permission of Ray Holt.

As a newly arrived engineering student at the Naval Postgraduate School, Monterey, CA, I had been fortuitously introduced to the elegant *SYM-1* microcomputer by Professor Russel Richards in 1980. My 1966 robots, *Walter* and *Crawler I*, were devoid of any type of sensors, controlled by a human operator via a multi-conductor tether. My first autonomous robot, *Crawler II*, featured four tactile sensors for collision avoidance, with a primitive punched-card reader to alter course accordingly. My primary thesis goal with *ROBART I* was to incorporate as many sensors as possible to enable far more intelligent autonomous behaviors. And significantly more sensors on a battery-powered robot naturally required a small, low-cost, but very capable computer with considerable input/output (I/O) capability.

For around $200, the *SY6532 RAM I/O Timer Array* and three *SY6522 Versatile Interface Adaptors* of the *SYM-1* provided an impressive 64 bits of I/O, which I further expanded with off-board 16-line data selectors and distributors. Buoyed by this seemingly limitless I/O and now a real computer to provide intelligent control, I set about to procure or build an appropriate array of sensors to effectively perceive the robot's environment. I'll never forget the thrill of accomplishment each time some innovative perception routine successfully enabled a new behavior, or how lucky I was that my obsession towards this goal was something the Navy was paying me to do.

Bart Everett with Robart I

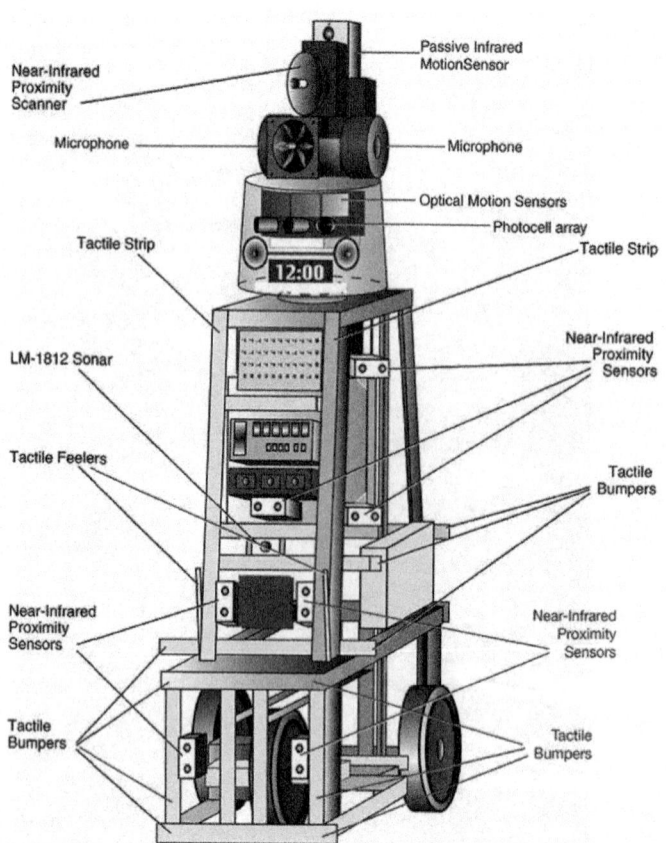

ROBART I, circa 1982, featured an extensive array of commercial and custom perception sensors.

As the first world's autonomous security robot, and to the best of my knowledge the first robot to incorporate speech synthesis, *ROBART I* naturally drew quite a bit of interest at the time. During the last year of my thesis work (1982), I was asked to give a presentation to the local Apple Computer Users Group. This limited exposure immediately generated several calls from local television stations, which came to my residence to film the talking robot that patrolled and protected my home. ABC World News Tonight flew out from New York to do a shoot, which started a world-wide media frenzy, and *ROBART I* went viral long before there was an internet. The Naval Postgraduate School sent their Public Affairs Officer out to my house for a week to answer the phone, which rang again as soon as you hung it up.

So, given the key role the landmark *SYM-1* played in my early 80s robotic experience, I had happily given Ray Holt an in-depth guided tour of our Lab's ongoing robotic projects in 2004, many of which were spawned by the supporting technology base initially developed under the *ROBART* series. Ray later arranged for a follow-up visit in 2006, accompanied this time with Manny Lemas, President of Synertek Systems, and programmer on the SYM-1, and Mark Holt, Ray's oldest son. During this second tour, I put particular emphasis on the perception and autonomy technology developed under the *ROBART* series, and its impressive legacy of literally hundreds of follow-on projects we subsequently produced for DoD customers. Without Ray Holt and the *SYM-1*, there likely would have been no *ROBART I*, whereupon I would have never been designated as Special Assistant for Robotics to the Commander, Naval Sea Systems Command, and then spent 34 more years working in the field of unmanned systems.

Foreword – Dave "Bio" Baranek – Former F-14 "Tomcat" RIO and Topgun instructor

When the Navy first flew its F-14 Tomcat fighter in 1970, it was the world's most advanced fighter and ushered in a whole new generation of tactical aircraft capabilities. This would be an impressive accomplishment for any aircraft, made even more so by the fact that the Tomcat was designed to take off and land on aircraft carriers. F-14 capabilities included a top speed of more than 1,500 mph, but also the ability to maneuver better than any existing or expected threat fighter, and a relatively slow speed for safer carrier landings. To achieve these conflicting goals designers chose a variable geometry wing, also known as a "swing wing."

Once again, operating a swing-wing fighter would be an impressive accomplishment, but the Tomcat's wings were computer-controlled, an aviation FIRST – and also a milestone in the history of electronics, since the wings were controlled by the Central Air Data Computer (CADC), which relied on a microprocessor. But there was no such thing as a microprocessor, so designers Ray Holt and Steve Geller INVENTED it. Their microprocessor was much more reliable than the electromechanical systems used by other aircraft, as well as being a fraction of their size. Thus equipped, the F-14 went on to long successful service in the US Navy, and its microprocessor-controlled CADC remained a key component. In a note that is sure to create some anxiety, the F-14 continues to serve (as of 2018) with the air force of one of our bitterest enemies, Iran.

Invented the microprocessor? So why don't you recognize the name Ray Holt? As co-designer of the world's first microprocessor he should be as famous as Edison, or at least Jobs and Wozniak. Unfortunately, the US Navy realized that this new capability was critical technology and clamped a lid of secrecy on it, which no one was able to remove until 1998. By that time the field was

©2006-22 Copyright by Ray Holt All Rights Reserved   2nd Edition
This publication cannot be reproduced in any form, without the written permission of Ray Holt.

crowded with names clamoring to be the first, leaving Holt and Geller to elbow their way to the goal for the recognition they deserved. (Unfortunately, Steve Geller has passed away.)

Ray Holt's book, *The Accidental Engineer*, tells this part of the story but doesn't dwell on it. No, in typical Ray Holt style, the story is relentlessly positive and optimistic, as well as very candid.

I thoroughly enjoyed the 1950s and 1960s that young Ray Holt knew, which I visited by reading his book. He describes a wholesome life. It definitely wasn't easy, as he dealt with setbacks, as well as uncertainty about where his life would lead. But his dogged persistence to succeed – at whatever task was in his sights – is evident on every page. As he himself admits, Ray was not the greatest student. Success wasn't handed to him, nor was he sure exactly what path his life would take. In fact, reading the book I wondered, "How did this guy come to develop such an important electronic device?"

That is one of the enjoyable aspects of Ray's life story. He kept working to succeed, all the while gathering the tools, he would need to accomplish this remarkable feat. Later in life he applies his optimism and persistence to another worthwhile cause: teaching robotics and science, technology, engineering and mathematics (STEM) to students in rural Mississippi. It's no surprise that this is also going well. Taking this journey with Ray, as told in his unassuming manner, was a memorable experience for me.

Thanks, Ray, for designing the F-14's CADC microprocessor … and also for taking the time to write your wonderful book.

September 2018
Dave "Bio" Baranek
Former F-14 RIO and Topgun instructor
Author: *Topgun Days*

Dave Baranek was born and raised in Jacksonville, Florida, and in his early teens he set his sights on flying Navy jet fighters. He attended Georgia Tech and participated in the ROTC to qualify for officer training, then entered the Navy in 1979.

Without 20/20 eyesight he could not become a pilot, so instead he became a radar intercept officer (RIO), operating the weapons system in the Navy's hot new F-14 Tomcat fighter. Shortly after he joined his first squadron, he received the callsign "Bio," which many of his former squadron mates still call him.

His first F-14 squadron was VF-24, and based on his outstanding performance he was selected to become an aerial combat instructor at the elite Navy Fighter Weapons School, better known as Topgun. While serving as an instructor in 1985, he had the unusual experience of flying aerial sequences used in the film "Top Gun," starring Tom Cruise and produced by Jerry Bruckheimer.

His second F-14 squadron was VF-2, after which he served in two shore tours: supporting the Joint Chiefs of Staff and the US 7th Fleet. He commanded F-14 squadron VF-211, responsible for nearly 300 people and 14 aircraft worth about $700 million.

He retired from the Navy in 1999 after a 20-year career. In retirement he has written several books about the F-14 and Topgun. His website is www.TopgunBio.com.

©2006-22 Copyright by Ray Holt All Rights Reserved   2nd Edition
This publication cannot be reproduced in any form, without the written permission of Ray Holt.

CHAPTER 1
Baseball, Radios, and Engineering

I was going to be a baseball player.

Baseball became the center of my life at a very young age, and I decided then that America's favorite pastime was going to remain as the center of my life forever. Anything "baseball" was better than anything else. I had played Little League since age eight, and I thought I was a pretty good player, so nothing would be stopping me from my baseball career. What I didn't realize was how life has a way of challenging and changing even the best-laid plans.

Little did I know in 1956, that at age 12, life would start changing my plans. During that year of my life, my next-door neighbor who was a radio technician was moving, and he asked me if I wanted some of his equipment. He took me into his garage and showed me what he had. Well, I wasn't too crazy about the radio stuff. Up to this point in my life, I had not enjoyed the concept of electricity too much. It was invisible, dangerous, and mysterious. It just wasn't something that I enjoyed thinking about or being around, but along with the radio equipment, he offered me his workbench chair which was tall and wooden with armrests. It was the best chair I had ever seen; however, to get it, I had to accept the radio equipment. So, I did.

©2006-22 Copyright by Ray Holt All Rights Reserved   2nd Edition
This publication cannot be reproduced in any form, without the written permission of Ray Holt.

The chair fit perfectly in my room, as my bed was tall and, by sitting in the chair, I could be at the height of my bed. Really, the chair was a very nice addition to my life. I loved that chair. It made me feel like a king.

My neighbor showed me a simple way to get radios working and how to care for them. He showed me how to open the cases; clean out the dust; and remove, clean, and replace the tubes before trying a radio. He also showed me how to take tubes that didn't light up to a local store to test and replace them if needed. With his teaching, I was able to get working 90% of all broken radios I attempted to fix. At 12 years old, I was actually repairing radios and getting some money from it. It was a great supplement to my paper route. I made about $100 a month from my paper route and about $30 more repairing radios. I also fixed bicycles for $5 to $10. Little did I know that this time with the radios would actually propel me into a full career. Had you asked me then if I would have a career in electronics, I would have laughed because as far as I was concerned all of my prime time and energy was for the baseball career I had planned.

Almost every summer from 8th – 12th grade was playing baseball. I started in Little League, then Pony League, Babe Ruth League then some private leagues. Rick Kiel, Joe Rodriguez, John Clausi, and Paul Heller were my most frequent baseball buddies. Rick and I played almost every day. During my junior high school days (7th-8th) we often play a game called Over-The-Line. It was a new game and even to this day I think we invented it. My Over-The-Line buddies were Rick Kiel, Chuck Kanoy, Ron Forsythe, Roy Clinton, and a few more whose names I have forgotten.

Nevertheless, I grew up, and so did my dreams. I still had not given up on becoming a professional baseball player, but I did begin to think about college and "regular" jobs too. In 1962, I was a senior at Dominguez High School in Compton, California. As part of our senior year, we had to take career and skills tests so the counselors could guide us into the right college or into some trade career. All I remember being told during that experience was that I "should not go into engineering" because I had a low mechanical aptitude. Mechanical engineering was the "big" engineering degree and the basis for anyone going into engineering. Some of my friends were actually accepted into great engineering colleges, like Cal Tech and UCLA, and it really disappointed me that, based on my test results, I would not be encouraged toward engineering. I felt I was just as capable as they were. Unfortunately, no one else told me so, and although many people told me what I should not do, no one told me what I *should* be doing. I did not know the college process. I even asked after high school graduation why I wasn't considered for some of the scholarships and I was told: *"you didn't apply."* I was lost.

As a result, I tried different things, fumbling my way through, trying to find my path. I had a desire to be an FBI agent, attended a career day conference, and talked to FBI agents at their booth. After talking with them a few moments, I was informed that anyone who wore glasses did not qualify because everyone had to be a field agent first and field agents were not allowed to wear glasses or anything else that made them unique. This is not true today, but as for my interest then, so much for being an FBI agent. Next, my father talked me into taking the entrance test for his welding and pipefitting profession. If I passed, I would be offered to attend some special schooling for the trade. I did pass with flying colors, and apparently, scored very high. I am sure my father was very proud of me. However, I had no motivation or desire to do what he did since all I could think about were his long and hard days, and neither appealed to me. He was very disappointed when I told him I was not interested. It was actually very hard to disappoint him considering we didn't get along that well anyways. My decision just made our relationship worse.

With no clear direction other than making something productive out of my life and no baseball recruiters at my door, my only choice left was to find some part-time job and attend the local community college, Compton Junior College. I found a job as an assistant leader of after-school recreation at the elementary school I had attended, Stephen Foster Elementary. It was not a difficult job, and it was actually fun. On the other hand, my first semester of college was quite difficult as I had no motivation. I took basic business classes and the normal English and history classes. Honestly, it was boring and very difficult to attend. Besides my part-time job, the only real fun I can remember having during that time was riding my new racing bicycle from home to college each day, probably a ten-mile trip each way.

In addition to my lack of direction and motivation and overall dissatisfaction with my life as it was, my home life was difficult because my father and I did not get along. I am sure he thought I was going to be some kind of failure and was just wasting my life. I would not admit it at the time, but as angry as he made me, I wasn't so sure he was wrong. Eventually, the constant fighting with him took its toll on me. I was so discouraged that I just walked out of college and quit my job. My decision only forced me to grow up that much faster. Now, I was forced to find a real, full-time job and make some money.

A great friend of mine, Bob Thornton, told me his father was looking for someone to work at his construction site. His father was also the local and district Little League director and knew me from my Little League days. He was willing to hire me and give me a chance although I was young and inexperienced for the work. The pay was minimum wage which was $1.75 an hour at that time. The construction site turned out to be a waste dump disposal site or garbage landfill. So, my first job was to stand knee deep in the garbage all day and water all of the trash so it could be compacted by the heavy equipment. It was the type of job you could perform without thinking so I had plenty of time to let my mind wander. From this experience, I have always said, "If you want to be motivated to get back to college, then, take a job watering garbage." All day long, among the hot, toxic fumes, I had lots of time to think about life choices and careers and cars and family and just about everything. The one thing I knew for sure was that I didn't want to water garbage for the rest of my life. Fortunately, after about six months, I was able to get a promotion to the gatekeeper position which was really a nice job. During the afternoon I would greet and collect money from the people entering to dump their trash and garbage, and then, after we closed, I still had about another four hours to myself, during which I usually read.

With new found motivation, *to not be a career garbage man*, and direction, *get a job that pays better*, I excitedly went back to college for the second semester of the current school year and was able to maintain my job at night. With enough time to study, getting paid, and a sense of purpose, life was actually looking pretty good. I was most excited about joining the college baseball team, and practice and tryouts began a few weeks after the college term started that school year. In the beginning, everything about my first college baseball experience was great. Many of my friends, including friends I had formerly competed against in high school, were trying out. As we got closer to the start of the college baseball season, the coach informed everyone that he would do an academic check and then decide who would stay. I did not know what all that involved, but soon I would learn the impact it would have on me and my baseball career.

My coach informed me that I was not eligible to play baseball that semester because I had all Fs on my last semester transcripts. I asked how that was possible since I did not even finish that semester. I learned that because I failed to 'officially' drop or withdraw from my classes I received an F in every class due to lack of attendance. I begged the college office to change my grades to reflect what really happened, but they said I had to prove myself with good grades before they would do that. So, there went my first college baseball season. Though disappointed, I was not too discouraged since I felt confident, I would make the team the next year, and afterward, continue my baseball career. Besides, a career in baseball was still the thing that I was most passionate about in my life.

Once I reentered college for the second semester, I still had NOT selected a major. I could not say I had any plans but I do remember saying I was a business major as I had some interest in business organization and corporate structures. I soon became bored with these classes as there was so much reading and I was not a good reader at that time. I changed majors at least four times, maybe more, business, accounting, forestry, engineering, later adding computer science and education.

Having a goal of "proving myself with good grades to college officials so that I could get my transcript changed so I could play baseball" in mind, I signed up for a class in trigonometry (Trig). I wanted to take this class because I felt I had something to prove that went back to a failure in high school. I failed Trig because despite being a student leader and past class president, I was kicked out of class for my attitude. I had a really bad attitude towards the Trig teacher for two reasons. First, as was my lifelong story, the problems between my father and I caused problems at home which made me angry at school. Secondly, my Trig teacher was always comparing me to my brother Bill, who was younger by two years and already known as a math genius. Unfairly, she (like everyone else whenever they met me) expected me to be just as smart in math. All the family problems and pressure of expectation worked to prove them all wrong in the end. So, retaking Trig in college was my way of proving to myself and everyone who doubted my abilities or witnessed my failures that I actually could succeed in math. I earned an A in Trig and went on to earn an A in slide rule, which were both pre-engineering classes. Ironically, despite being told in high school that I should not major in engineering, I was doing well in my college engineering classes. I was surprised at the time by my unexpected success in this area, but every success gave me a little more encouragement to keep going academically.

Aside from school and work, I kept busy by playing on a local baseball team, the Compton Travelers. We played at Cressey Park, and I enjoyed everything about the experience. There were several talented, local players on the team with over half of the team going on to play professionally: Jim Rooker, Lynde Kurt, Rick Kiel, Reggie Smith, Roy White, Mike Paul, and Don Wilson. It was during this summer that I unknowingly took yet another step closer to my future career. My friend Joe Rodriguez was home visiting, having returned from the University of Idaho where he was playing football. He really pushed me to come with him the next year, saying we could room together, and giving me a catalog to look through. The beauty of the campus in the pictures is what made my decision. I had never seen a college with so many trees, and I was really captivated. The next day I asked my parents if I could transfer, and they said I could go if I got accepted. Well, it may have seemed like things were going well but I was in transition for sure. The thought of attending and graduating from such a big university was very exciting and, of course, the thought of leaving home and being on my own was a huge factor. No one in my family had gone to college so it was a big step for all of us.

I was concerned about the Fs on my transcript as I filled out the application, but I also thought about my recent A's. I was concerned that the combination of me being an out-of-state student combined with the failing grades on my transcript might keep me from being accepted, but I applied anyway. The application process also forced me to do something I had not done before up to this point in my college career, select a major course of study. I did not feel that I had anything to base my selection on other than I knew, based on what I had been told throughout my high school career, engineering was not a major I should consider. After looking through the catalog of courses offered, I decided on forestry. Forestry seemed to fit right in with my desire to be outdoors and with my interest in conservation. I actually was getting a little excited, and after receiving an acceptance letter two weeks later, I finally felt like I was going in the right direction with my life.

Many decades later, somebody asked me a question about my choice: how DID you get accepted at Idaho given your academic record? That is a question I can't answer. I did not expect to get accepted. All I can think of is that the school had a quota for out-of-state students and they wanted to give me a chance. The last semester of grades before I applied were not too bad and showed I had some ability. I think of this often and usually conclude that God just opened this door for me. Being 1,200 miles away allowed me to become the person of my own choosing and not that I thought others required of me. It took a while to happen and many events during these colleges' days shaped my decisions .... including new friends, girls, Army ROTC, Judo classes, and the small-town atmosphere.

1963 Army ROTC weekend war game. Ray on left.

During that next month, I saved as much money as I could and prepared myself to leave home for college. This was about to be by far the biggest change I had experienced in my life. It was 1,200 miles from home in Compton, California, in Los Angeles to the University of Idaho in Moscow, Idaho. This was not just a move to college but an actual move from home. I think my parents felt it more than I did. On one hand, I was eager to get out and prove myself so I would not fight with my father so much about me making something out of my life, on the other hand, I really had no idea what I was doing, everything was so new to me. In my preparation to go, I remember my parents buying me a travel trunk for all of my clothes. The trunk made me feel like an official college student since at the time that was the usual way college students carried their belongings to school. I still have that trunk today, and although it's in really bad shape, it contains many college and high school memories.

My trip to the University of Idaho was by Greyhound bus. It would be my first bus trip, and I did not know or think about how long it would take. I was just excited to go. My father took me to the bus station, and as I was about to get on the bus, he extended his hand, shook mine, and told me to "do good." That was the first time I remember him shaking hands with me or giving me any encouragement. It gave me a great feeling, but it also had the effect of making me want to stay and see if maybe things could actually get better between us. However, the bus was ready to go, and I knew I needed to leave with it. That bus trip was the longest trip I had ever taken. It took 12 hours to get to Boise, Idaho, which was in the southern part of the state. I thought I had almost reached my destination at that point, but it took another 10 hours to get to Moscow, Idaho, which was in the northern part of the state. I was exhausted by the time I arrived. I was so far from home that I felt I would never go back.

University of Idaho, Moscow, ID

When I did finally arrive, I was pleased to see the college campus was as pretty in person as the pictures in the catalog. I felt like I was in another country. It was a great feeling to be on my own. I was finally making decisions about my future and seeing them through. I had a plan to finish college with a degree in forestry. However, there was no deep purpose or conviction behind any of my decisions. I was moving forward, and I was satisfied with that. My first year at the University of Idaho had its ups and down. I was a part of the Army Reserve Officers' Training Corps (ROTC) which taught me discipline, something I really needed at the time, and I received many awards and honors, such as carrying the American flag for the unit. I also made a lot of great friends: Bill Donnell, Bill Roper, Charlie Beyer, George Nipp, and Brian

Stickney to name a few, which helped build my self-esteem. These were a few of my ups.

As for my downs, my first year was an academic disaster. I had to take English, history, American Government, botany, and forestry, and I barely passed my classes. Another disappointment was baseball. This was probably the only thing I had ever felt a strong sense of passion, purpose, or conviction about my entire life. Though I had made little progress in achieving my dream, I was still hanging on to my desire to play professional baseball. I tried out for the university team and thought I was doing quite well against the older players. Right before the season started, I was told I had to get a physical exam from a local doctor. The doctor discovered during the exam that I had a torn muscle. I had to have surgery, and it was not possible to recover and still make the team. I knew this was the end of my professional baseball career hopes. This realization was very discouraging and almost put me in a major depressed state. My surgery was scheduled during spring break. I was not able to go home, and my parents were not able to afford to come see me. I spent a week in the hospital very lonely and in a low mood. Without my professional baseball career dream, I felt lost, and I began to think the only option I had now was to get my forestry degree. With my less than acceptable academic progress, I was not even sure that would happen.

With no other or better options in sight, I returned to the University of Idaho for a second year in spite of my disastrous first year. My academic major had some potential, and it was the only thing that made sense in my life at the time. I joined the Forestry Club and met another good friend, Lyn Thaldorf. Lyn was a born, forestry-type of guy. He loved the work, eventually had a very successful career, and today is on the Board of Trustees for the Forestry

College. Lyn encouraged me in many forestry activities, including the Forestry Club retreat where I won the 2nd place in an axe-throwing contest. Lyn and I worked the next summer together in the Coeur d'Alene National Forest in Northern Idaho. I had the job of Recreational Director, or Ranger Ray, as the kids would call me. During the day, I would meet and greet the families, and on weekends, I would clean toilets. A very humble and yet rewarding summer. The off-days of fly fishing in the Coeur d'Alene River, hiking, and driving to the lookout towers were exhilarating. I was content although not completely happy with my life. To add to my heartbreak, it seemed baseball was destined to be a part of my life even if I could never be a part of the game.

While I was not following his progress in detail, my brother Bill had graduated from high school by this time and received a baseball scholarship to Stanford University. He actually turned down an academic scholarship in math because he thought baseball would be a better path toward graduation. As with most things he set out to do, he did quite well. He was an outfielder and pitcher when Stanford participated in their first College World Series in Omaha, NB. I was really proud of him. My sister, Elizabeth, affectionately called Betty, and later Liz, married her high school sweetheart, Jim Rooker, after high school. Jim went on to become a professional baseball player for the Pittsburgh Pirates. Liz and Jim spent their lives in professional baseball. Jim was a success on the field, and Liz became a success in her own right as a baseball player's wife. Many have said that she was as well-known as Jim. Their career highlight was when Pittsburgh won the Baseball World Series with Jim pitching a key game. I was happy for my family, and I knew that baseball was not my future. Yet, choosing to let baseball go was not an easy choice, but it was necessary.

My second year at the University of Idaho was a year of maturing. Instead of pursuing varsity baseball, I decided to take Judo lessons from a local policeman. I progressed three belts that school year, entered my first and only competition in Portland, Oregon, and won first place over competitors from Northern California, Washington, Oregon, and Idaho. I came close to earning my brown belt in Judo, but I injured myself a few days before the test and never did try again, something I still regret. I also played intramural softball. The intramural softball season climaxed with a final game between my independent league team and the fraternity boys, Sigma Alpha Chi (SAC), who had not been beaten in any recent years. Two sorority girls, DJ Green and Carolyn Brown, probably the only sorority girls cheering for us, were among those pulling for our team. In the end, we did win, partially from me hitting both left and right-handed home runs. The victory seemed to be what I finally needed to release my thoughts of a professional baseball career and move forward. I guess I finally realized I was as good a player as I needed to be, and I had better start planning for another career.

Going back to my career plan, after taking the Physics of Electricity class and doing perfectly on homework, test, and exams I realized that I could really do it. Also, that same semester I received an A in Calculus just because I worked hard studying and worked every problem in the book. It was hard work but it paid off with nice grades. The best semester ever in college. I made the Dean's list.

The physics class satisfied a physical science requirement for forestry so I did not think about it being engineering. The Calculus was required for forestry so again I did not think about it being for engineering. However, when I started looking for another major degree program, I soon realized that I was actually taking engineering classes and doing well. That was a huge motivation.

There is also a motivation that pushes one to go further. I think my motivation was "I am getting older and not accomplishing much." I think next was the fact that I realized that I needed to take control of my life and not let others decide for me.

About this time, the turning point in my life that would firmly plant my feet onto my lifelong career path occurred. To be considered a serious forestry student, I needed to maintain an A or B in chemistry. Yet, I had a very difficult time in chemistry since I had never taken it in high school, like so many others in my class. I could not grasp the concepts fast enough. Dean Ernest Wohletz of the Forestry Department, he was about 70 years old with a desk made out of the cross-section of a very large tree, called me in one day near the end of the semester and asked me if I liked forestry. I told him I did.

Dean Wohletz said if I wanted to continue in forestry, I would have to do better in chemistry. After a further discussion that I don't remember too well, he said I "should be taking this class" and pointed to a particular class in the school course catalog. To this day, I do not know how he decided to recommend the class to me. It must have been academic wisdom. Whatever it was, I am forever grateful as the outcome of the class changed my college, life, career for the good.

CHAPTER 2

The Accidental Engineer

The class Dean Wohletz recommended was physics of electricity. I was not too thrilled but willing to take it because I remembered how well I was able to fix radios and how much I enjoyed it. I soon learned the class had nothing to do with fixing radios, but I did like the class and got all As on every homework assignment, test, and the final. This was the first time in my life this happened and it was in a class that was a part of a pre-engineering major, electrical engineering. The entire experience made me think. I would have never said I was fond of electricity before, but maybe I did like electricity after all. I was always told I should not go into engineering before, but maybe I could do this. I was not sure where my new found realizations were going to take me, but I was ready to take on another course and do as well. The same semester I took Calculus and made an A. My academic motivation was higher than it had ever been. I did so well that I actually made the Dean's List for the first time in my three-year college career.

This accomplishment was first told to my parents in a nice letter from the university. Upon receiving the letter, my mother immediately called me and said she thought a mistake was made. I proudly told her it was no mistake and that I was really doing well. I think for the first time in my life my parents thought I might truly accomplish something. I was happy to make them proud. I felt like their sacrifices were finally being rewarded. My mother had taken on a second job to supplement my father's income so my brother and I could attend college. To this day, I have no idea how they supported all of my expenses, my brother's expenses outside of his scholarship, and my sister's expenses. To this day, it all remains a mystery to me, but I have a memory of great appreciation for the entire experience of my parents' financial sacrifice.

As happy as I was, I was disappointed that I had spent so much of what I felt was wasted time when I could have been on this path all along; however, I accepted that this was the way it was supposed to happen for me. This was the way I was supposed to get to where I needed to be, and I picked up some good friends and experiences along the way. With a new motivation and direction, I began to look into what choices I had. I began researching colleges that had electrical engineering degrees. The University of Idaho did obviously have an Electrical Engineering degree, but I was motivated to search all my options. If I had stayed at the University of Idaho that would have been ok. My college search ended when I found an electronic engineering program at California Polytechnic University (Cal Poly) just outside of Los Angeles in Pomona, California. I was attracted to this college because of the huge emphasis on "hands-on" training. I enjoyed "doing" as I learned. Before I could apply to and be accepted to Cal Poly, I needed to get over one last hurdle. I had to go back to Compton College and get them to change my F's to WD's or Withdrawals. At last, I applied and was accepted into Cal Poly.

The electronic engineering program was both very rigorous and structured. It would require me to go three more years as a full-time student to graduate, and that was only if I passed all of my classes and did not have to repeat any. The electronics program was new and a little different than the electrical program in that electronics would teach more about transistors, the new, very small semiconductor devices that were going to replace vacuum tubes; moreover, the electrical program taught more about motors and large power systems. I liked the idea of the new program and dealing with smaller voltages, around five volts instead of 1000 volts or more in electrical. I found out that a friend from my high school wrestling team, Jim Hooks, was also attending Cal Poly. Jim and I had a very nice relationship and often worked the many labs together and encouraged each other. Jim's presence and encouragement were a huge part of why I did so well while at Cal Poly. In three years at Cal Poly, I figure I took 25 labs and wrote over 200 lab reports. There was no doubt I was well prepared for the real world of electronics. There was nothing I feared in electronics, that is, except designing amplifiers.

In the middle of my Cal Poly education, I married a very nice girl named Lynda. She attended my high school but I did not know her at that time. Her brother introduced us later. We have three sons, Mark, Michael, and Brett. Lynda was a very accomplished elementary teacher. She was very supportive of my degree and found a banking job to help our income while she also went to college. I was able to get a job at AeroJet General, first as a tester of electronic torpedoes, the U.S. Navy Mark IV, a computerized self-guided torpedo, and later as an electronics incoming inspector. This inspector job proved to be extremely valuable as it allowed me to learn all about the practical aspects of components and how to test them. Between the two of us going to college and working, we did not see each other much. We were able to find a place for $30 a month, which greatly helped out. My parents had given us $500 for a wedding present, and that was a tremendous help in getting started. We had our first child, Mark, during my last year at Cal Poly, and even though it seemed like we had no time left, it still worked out. I would go to school from 8 am to 4 pm, work from 4 pm to midnight, and when I got home, Mark and I would see each other for a few hours while I did my homework. This became such a routine for Mark that even today as an adult he keeps similar hours.

Cal Poly Pomona, Pomona CA

During my last year at Cal Poly, I needed to take one more elective. It was suggested to me to consider a class in theory of switching systems. I was not too sure what this was, but it sounded interesting, and since I had previously taken a humanities class in logical thinking, and did well, I thought this might complement that type of learning. Well, it turned out to be a very pivotal class in my career. The class actually taught the theory of logic design, which is the foundational knowledge in computer design. At the time, computer design was not taught as a separate class. This class was the beginning of what would become Cal Poly's future computer design curriculum. Cal Poly taught computer programming but not the actual methods of designing the hardware.

There are two classes that really teach computer design: Computer Architecture and Computer Structures. Both classes go into the organization of a computer. The class I took was mainly a math class that dealt with the arithmetic of computers. I guess it could be said to be a computer design class, but since no other computer design class was taught, I always considered it a math class. Of course, I did not think too deeply about it all then. I needed an elective; I took the class.

If you are asking, well yes, it was a hands-on class. One hour of lecture a week and three hours lab. We designed and built electronic logic circuits. This became the fundamental experience I needed in order to succeed with the F-14 microprocessor. This was one of the most enjoyable classes I ever took.

I don't remember any reaction from my parents, so I wondered about it. I suspect they liked the fact I would be closer to home (about one hour). I have NO idea how they paid for the school and I was so selfish I don't even remember thinking about it. I learned later my Mom had two jobs and some work at home to support my brother and me. Even today I feel really bad I was so selfish.

I also had a brother, Bill, and a sister, Liz, as I mentioned earlier. For some reason, I just did not communicate with them. There was no problem between us:

William "Bill" Holt

©2006-22 Copyright by Ray Holt All Rights Reserved   2nd Edition
This publication cannot be reproduced in any form, without the written permission of Ray Holt.

I guess it was just because communicating was a little more expensive and time-consuming, so I just didn't take the time.

Graduation day finally came. I remember it well as my father and mother both attended, and I could finally show them that I made it and all of their hard work paid off. I was relieved that I proved I wasn't a failure. How could I be a failure with an electronic engineering degree? I was proud of my accomplishment, but I learned that education, if truth be told, starts the day you graduate. During this time in history, technology was changing every 1-2 years. That meant as I was earning my degree, at least two changes in technology had occurred. For someone seeking employment in a technology field, this is a huge reality.

This point has to be stressed: technology studies are totally different to any other field of knowledge, mainly because of its continuous acceleration. Thankfully, while I was in my last semester, the engineering college organized career interview days with various companies interested in Cal Poly students. We were in demand because of our hands-on experience. Most, if not all, graduates were offered great jobs. The aerospace industry was rapidly growing. The Vietnam War was escalating. I remember getting three job offers: Westinghouse, working on space designs; Bendix, working with torpedoes; and Garrett AiResearch, working on aircraft design. I actually liked the Westinghouse job best as it was in logic design, and I liked the Garrett job the least as I was told I would be designing amplifiers which I did not like at all. Garrett wanted me to design amplifiers, a kind of circuitry I never liked. But I accepted the job. I sometimes try to imagine my life if I chose Bendix, or Westinghouse: space was, and still is, the final frontier.

Coincidentally, or maybe it was fate, my brother graduated from Stanford University the week before me and was also interviewing with Garrett as a systems programmer. I always say I took the 6-year college plan and Bill the 4-year college plan. We talked about Garrett, but because the jobs were different and in different divisions, we really didn't expect to see each other much "if" we were both hired. Though not my first choice, Garrett was very aggressive and consistent in recruiting me, eventually making me a very nice offer. In the end, I decided to take their offer. Bill, in a very independent decision, also decided to work for Garrett. It was nice to know we would be working for the same company, but again, since our work was in different areas, hardware design and software programming, we did not expect to work together. Our work buildings were about 10 miles apart.

The first day I walked into Garrett I was met by the personnel manager, the same as a human resource manager today. His name was Dick Gentry. Dick took me into his office and asked me to sit down. He proceeded to explain some company benefits and had me sign the usual papers. What he did next was the final step that would take me the rest of the way down the path to my destiny, to what my life had already been

Ray at Cal Poly Pomona Graduation

carved to be. Before I sat down with Dick, life had been nudging me first toward and then down a path that I couldn't see, but Dick's next words firmly and finally ended my journey and announced my destiny by revealing my future.

Dick opened up my personal folder and asked, "I see you have taken a computer class?" I was a little shocked because I really didn't consider it a computer class, and so he had to remind me of the switching theory class. Very nervously I finally said, "Yes." His next statement worried me even more. He said, "You are the only one in your department to have a formal computer design class." I had no idea where he was going with this, but I knew one class does not make a computer designer. I could not imagine where his questions were leading since, I was hired to do amplifier design, and I had studied extra the month before just to make sure I could do that! Then, he got up and told me to follow him. He took me downstairs into the basement of the building.

As I remember, this was a huge room full of table areas with equipment and people who were working on everything. It was a very organized but busy place. He walked me over to this very large piece of equipment, took off the top piece, and asked me to look inside. He asked, "Do you know what this is?" It looked like an oversized transmission. I knew because of its size it was not a normal

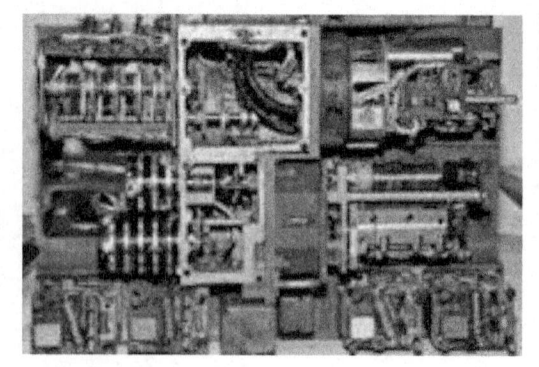

F-4 Phantom Jet Mechanical Flight Computer

©2006-22 Copyright by Ray Holt All Rights Reserved   2nd Edition
This publication cannot be reproduced in any form, without the written permission of Ray Holt.

transmission. In addition to being larger than normal, the inside was full of chrome gears, cams, and electronics. A cam is an irregular shape on a shaft. It can be circular or it can be oblong or any shape in between. In a mechanical computer, it is usually based on some mathematical equation.

I knew I was wrong, but without any better ideas, I told him my first guess.

"It looks like a transmission."

"No," he said, "it is a flight computer for the F-4 Phantom Jet." The F-4 Phantom Jet was THE jet flying in the Vietnam War. Included photo is exactly what I was shown.

I remember thinking, "So, what?" At this point, I still didn't understand how and when we got onto this new topic of conversation. To be quite honest, at this point, I could no longer even identify the topic of conversation.

Then, Dick said, "We want you to work on a special project that will convert this mechanical computer to a 100% electronic computer for a new airplane."

...Did he just say he wanted me to DESIGN an entire computer??

CHAPTER 3
The Navy Calls—The F-14 Brain

My first week at my new job was really slow. I found out that the company had not received the contract yet but they were confident they were going to get it. I felt like I could be fired any day if they did not get it. The work environment was very much what was called an "aerospace bullpen" arrangement. This is where most of the workers are at desks in one big area with few walls and the bosses are in offices on the side. My desk was in a smaller area with my immediate boss, Dick Barcus and our documentation expert, Larry Hammond. Larry and his wife Julie became good friends but we lost contact about ten years later.

There were about 50 engineers in the area with about 8 senior advisors, and I was the last team member to join. Garrett AiResearch was working on several projects. The other logic designer, Steve Geller, was also just hired to work on the computer. Steve and I were to work together. Steve was about 45 years old and had computer design experience. At first, I was very much intimidated by Steve. You might say Steve was a typical computer geek. He did not like to talk about unimportant things and he liked to be left alone when he was thinking about his computer designs. The very unfortunate part was that Steve's desk was right in the walkway where everyone bothered him. I knew it was nerve-racking for him. I think for the important role he was asked to play that he deserved an office with a door he could close.

Steve and I did not get along at first. I think he was frustrated with me since I was new and had little experience. I was always asking questions. Later when he found out that I liked to "touch" electronics, we got along very well. Steve did not like electricity and would not go close to anything that really worked. In fact, it took him about a year to actually see our computer prototype working. Steve was very much a "high-level" computer designer. This is normally called a *systems designer*. He would come up with high-level (general, not detailed) design ideas and would explain them to me to see what I thought. Because I was a new computer designer it was hard to follow his ideas. I did much better and began to contribute to his thinking after a few months. Eventually, Steve and I became really good design partners. He would work on the high-level ideas and I would do the actual detail design and make it work. After the project was completed, I never did see Steve again and I do not know what happened to him. His contribution made the project possible and successful. He was a very unique person and perfect for the job.

I did most of my hands-on work with Lynn Hawkins. Lynn was the Senior Laboratory Electronics Technician. Lynn was very talented in taking paper design ideas, building them, and proving that they worked or not. Lynn and I become very good friends and stay in touch today. Lynn's skill highly contributed to the success of the computer being on time and working the first time.

The project was called the CADC which is an acronym for Central Air Data Computer. This computer was for a new airplane for the United States Navy. I learned later the pilots just call it the ADC. The details of the airplane were secret and I do not think too many people knew everything about the airplane and what it could do. Garrett AiResearch was given the contract to design and build the CADC. The airplane was eventually named the F-14 and later given the nickname "Tom Cat." A famous movie, called "Top Gun", was made featuring the "Tom Cat" and recently another moved, called Maverick. Both movies starred Tom Cruise.

The CADC was to consist of two pressure sensors, a computer with the ability to give commands to several movable surfaces including the wings, flight data to the pilot and the weapons system, communications data, some interface electronics, and a power unit. My (and Steve and Lynn's) job was to design and build the computer. Later I will talk about the others that helped make the computer successful. The entire team for the CADC consisted of 25 engineers and scientists, including the bosses. Others worked on the pressure sensor, the box design, the connection to the moveable wing, and the power unit. Tom Redfern was in charge of engineers that did most of this. Tom and I stayed in touch for many years.

Tom Redfern. "I have not thought of my work on the CADC in years. I am not sure exactly what is classified and what is not at this point. In particular, the pressure sensor was very advanced for its time. I remember a good deal about its construction and calibration.

As I recall the static air measurement was the only one that required beyond twelve bits of resolution. This was because at 80,000 feet there is very little pressure. In those days, the 1 ft. resolution was required by the weapons delivery system. When dropping a bomb, the plane was put in 'altitude hold' mode. The bombs where ballistic and any vertical velocity would affect the target accuracy. They could not use the radar altimeter because of ground variation. At 80,000 ft. air pressure is very uniform. This is all moot today because of smart bomb technology. I think the reason the CADC had such a high security classification was because of its interaction with the weapons delivery system.

The first air data computer I worked on was for the F-4 Phantom. It was electro-mechanical, built with gears, motors servo amplifiers and potentiometers. The potentiometers were wire wound and about 2 1/2 in. in diameter. The altitude function was linearized by varying the pitch of the wire on a potentiometer. Next, I worked on the CADC for the F-111. It was all solid state, mostly analog but had some digital functionality. The F-111 was a low-level bomber so it did not have the extreme altitude requirements of the F-14. Both the F-4 and F-111 used a force balance mechanical pressure sensor. The sensor consisted of an evacuated 'bellows' on one side and a voice coil on the other side. The sensor was in a servo loop and would never move. Pressure was proportion to the current in the voice coil. Since the balance beam never physically moved the pressure sensor had very high frequency response. This same sensor was used for the inlet control system on the SR-71. I only mention all this because of the rapid technology change from the F-4 to the F-14. As I look back it still amazes me." **(text by Tom Redfern)**

It took about three months for Garrett AiResearch to finally receive the contract. During those three months I was told to learn all I can about designing computers. Every day during those three months I would read and learn about different techniques on computer design. I was particularly interested in how a computer was able to add, subtract, multiply and divide so fast and accurately. Eventually, I became an expert in this area. It was my specialty area of computer design. Garrett AiResearch sent me to a week-long computer class at UCLA on designing computer arithmetic units. The teacher was very famous, Professor Gerald Estrin I really enjoyed the class. It gave me some very good ideas on designing computer arithmetic units.

As mentioned above, before this project the old way of designing computers for military airplanes was called a "mechanical" computer. It was mostly made from gears, cams (funny mechanical shaped pieces) and some electrical parts. Mechanical computers were very heavy and were very difficult to change after they were completed. This new CADC was to be made from electronic parts which would make it much lighter in weight and would allow for easier changes. Steve and I did not know exactly how we were going to do this new design. For now, we did not know the exact details. All we could do was to guess based on previous mechanical computer requirements. As far as I know, the Navy skipped the transistor-control phase, passing from mechanical devices to integrated chip devices.

The project was started in June of 1968 and completed in June of 1970. The first flight of the airplane was on December 21, 1970. The computer box or project was called the Central Air Data Computer or CADC. Some of the design constraints that was put upon us was that the computer electronics had to be within 40 square inches or less and that it was to consume no more than 10 watts of power and the cost of electronics had to be between $3,000 and $5,000 and the entire CADC box had to operate over what is called mil spec temperature range and that was from -55 degrees C to +125 degrees C. The CADC also

©2006-22 Copyright by Ray Holt All Rights Reserved   2nd Edition
This publication cannot be reproduced in any form, without the written permission of Ray Holt.

provided data for controlling and firing the six Phoenix and Sidewinder missiles at the same time. Other design constraints were acceleration, mechanical shock, reliability, and project scheduling.

Today there are several excellent YouTube videos that show the F-14 inflight and it is possible to observe the movement of the wings and of the other computer surfaces. You can also see the F-14 going through the sound barrier which is called Mach 1, occurring in the Earth's atmosphere when an object flies over the speed of about 767 miles per hour (343 meters per second).

Let's talk about the CADC and what it did. It basically was the flight computer for the F-14. It computed and displayed for the pilot the altitude, air speed, vertical speed, Mach number and temperature. The CADC also computed and controlled the wing sweep position and rate of movement. It also controlled the maneuver flaps which allowed the plane to rotate and the glove vane which was a moving surface on the front of the wings to stabilize the F-14 at high speeds. The CADC also computed the angle of attack which was important in landing and in firing missiles. The CADC provided critical information in real-time to two other systems on the airplane. These were the weapons systems and the communication systems. Both of those systems were also computers. The CADC also computed inflight self-test diagnostics which I will talk about later and it also had a duplicated switch-over dual computer system, called dual redundancy.

I would like to give an overview description of the CADC and how it worked. It was a digital computer system operating with one's and zero's that had to interface to a mechanical world. The mechanical world did not use ones and zeros. It had voltages that varied continuously so in order for the CADC to interface with the mechanical world it had to contain converters on the input of the CADC, the converters were called analog to digital converters. For the output of the CADC, the converters were called digital to analog converters. Our designers of these two converters pushed the technology approximately 8 to 16 times beyond what it was capable of in order to perform the accuracy needed for the F-14. After the mechanical movements were converted to digital positions then the computer performed the necessary calculations in order to fly the airplane. When the computer was required to move mechanical surfaces such as the wings, maneuver flaps, glove vanes, and the instruments on the pilot dashboard it would use the digital to analog converters. Another major important item in the CADC was the quartz pressure sensor. The main input to the CADC concerning the airplane's flight was from two pressure readings on the outside of the airplane. This pressure was then channeled into the CADC and to the quartz sensors and the quartz sensors would then measure the amount of pressure and converted it to a digital number and then that was sent to the computer. So, the entire CADC box had three areas of extreme state of the art technology; the computer, the analog to digital and digital to analog converters, and the quartz pressure sensor.

F-14 Static Pressure Sensor

Tom Redfern "Because of the resolution required, 1 foot in altitude at 80,000 feet, a digital pressure sensor was needed because no A/D converter was good enough at that time. The sensor was basically an inductor and capacitor connected in series. These elements formed a tank circuit that set an oscillators frequency. The capacitance varied with pressure causing the frequency of oscillation to vary with pressure. It was then a simple matter of measuring the frequency and hence air pressure.

It sounds simple but it is far from simple. The sensor must be repeatable and must be stable over the military temperature range (-55C - 125C). That means the inductor and capacitor must be independent of temperature. In addition, the pressure vs altitude function is extremely nonlinear and the capacitance vs pressure is nonlinear as well.

To solve these problems the sensor was constructed from quartz, because of its dimensional stability with temperature. The sensor consisted of two pieces. The bottom piece was a cylindrical tube with a helical square notch cut into the inside of this tube. In this notch a square copper wire, that fit snugly, was inserted. This formed and air core inductor that physically could not move because it was held in place by the quartz. Since there is virtually no expansion or contraction of the quartz over the military temperature range the inductance would not be temperature dependent. The top of the tube was closed and the circular center was plated with a metal coating.

The top piece of the sensor consisted of a quartz diaphragm plated with a metal coating. During the manufacturing process the diaphragm was sealed to the top of the tube in a vacuum. This formed a parallel plate capacitor that would vary with pressure as the diaphragm flexed. Thus, oscillating frequency would be a function of pressure.

Nothing was linear. The pressure vs frequent function as well as the pressure vs altitude function were both nonlinear. AiResearch had equipment that could very precisely generate pressure. Pressures were varied over the operating range and the pressure vs frequency function was determined. This provided the data required to linearize the sensor. A bipolar fusible link ROM was used to then linearize both the pressure vs frequency function and the pressure vs altitude function.

I don't know how it was done, but to build this sensor the quartz had to be machined and welded when the top piece was attached to the bottom tube."
(Text by Tom Redfern)

The design phase of the CADC and its microprocessor took about a year. The internal components for the CADC were all made with new technology. It was difficult to know at any one time if everything would fit. The microprocessor was one of the most unknown because we did not know how many chips it would take to make it work. I remember doing at least three designs of the microprocessor. Each design had to be evaluated for performance, size, cost, and feasibility. Included in all of this were the chip sizes, the power usage, and speed. Also, we had to consider that the CADC had to operate at military requirement of extreme temperature. This was a huge requirement to meet with new technology chips. Meeting this requirement alone was amazing.

Along with the above, the logic design and programming of the microprocessor had to perform many mathematical operations to fly the airplane. In order to prove this worked we built a hardware prototype out of standard circuits and we wrote a software simulator which checked my programming of the math. Mr James Lallas wrote this software simulator and did an excellent job of helping find any issues and flaws in the programming. Mr Lallas continued his career as an engineering professor at City College of San Francisco.

One of the functions the microprocessor had to perform was in-flight diagnostics. This means that the microprocessor had to test itself for failures and when it had detected that part of itself had failed then it was to notify external electronics. The reason we had to perform in-flight diagnostics was because about half way through the project the U.S. Navy asked us to provide the MTBF, or Mean-Time-Between-Failure for the design. The MTBF was the Navy's (or entire military) way of determining the reliability of the design. With the microprocessor using new technology, there was no data that represented as reliable design. This did not set well with the Navy and I remember the project being put on hold for a few weeks. During those weeks, I was asked if it were possible for the microprocessor chips to test themselves in-flight. This was a new area for me and after considering various means of self-test I decided it was possible. The Navy requested 100% self-test of any single failure inside the microchips. This meant that if any inside wire (or metal) shorted or opened or if any of the 70,000+ transistors failed that somehow, I had to make sure the failure was detected.

The decision that we could do the self-test was completely unfounded as I, nor anyone else, had ever done something like this before. Self-testing was not a new event at that time, but to be able to 100% detect failures in large and complex circuits were new and very complex tasks. In fact, the tasks were so large and complex that we had to use large business computers to simulate the process and help us make sure the results were accurate.

To perform this new design challenge, I asked for more programming help. My brother Bill, 2 years younger than me, already worked for AiResearch and had just finished designing a directed-beam helicopter search light and was available for the project. He was assigned to work with me on these diagnostic tasks. He was a scientific programmer using the companies IBM business computers. He used the Fortran IV computer language using punched cards as a stored program memory system. Fortran is a so-called high-level computer programming language. It was originally meant to FORmula TRANslation, i.e., math problems, but can be used for other purposes too.

After much discussion, Bill and I decided to simulate each of the microprocessor chips down to their transistor design level. This meant that he would write a program in FORTRAN that would simulate the exact logical workings of every circuit on every chip. This was not the normal intended use of Fortran but he was able to do it. The task was huge and seemingly impossible because of the complexity of chip details. I would estimate it took him at least three months to get the program ready. While he was working on the program, I was working on combinations of binary number patterns instructions that would exercise every wire and all of the logical circuits on the chips. Our intention was to have the microprocessor perform normal mathematical calculations and then compare the results to what we knew as a known result and if the calculated answer was different than the known result then there must have been a failure on one of the chips. I remember the last few days of this process and we put in many, many, very, very long hours and lots and lots of testing of number patterns to see if we could, in fact, detect any single failure. To prove that we could, Bill wrote the program in such a way that it would automatically go through and force simulated failures throughout the "chip simulation" and then see if we could detect a different value when compared with the real value thus detecting the failure if they are different.

The final conclusion was that we could 100% check all single failures on the connections and the transistors inside the microchips and 100% of all the

©2006-22 Copyright by Ray Holt All Rights Reserved   2nd Edition
This publication cannot be reproduced in any form, without the written permission of Ray Holt.

binary ones and zeros in the read only memory chip which contained instructions and data and 100% of the non-arithmetic chips. On the two arithmetic chips, the parallel multiplier and divider, we were able to detect 98% of all single failures. The 2% of single failures we were not able to detect were abnormal conditions of the F-14 and so the Navy approved our in-flight diagnostic solution over having no MTBF historical failure data on the microchips.

The Navy also asked us to duplicate the entire microprocessor, called dual redundancy, and provide a means of switching to the second microprocessor when the first microprocessor failed. We were also asked to provide a means of turning on a red light on the pilot dashboard when there was a failure. This light would be called Pilot Notification. It would tell the pilot that one of two computers had failed and that they were operating on the last computer. If both computers failed it would have been nearly impossible to control the F-14.

Bill and I stayed on the F-14 project about one year after the first units were delivered. We spent that time working on the Technical Manual that explained to others how the microprocessor worked and was programmed. Near the end of that year Bill physically passed out while programming on the key punch machine. He was rushed to the hospital and was diagnosed with a brain tumor. It was a new type of cancer and it had already reached the size of a baseball in his brain. The doctors could do nothing for him and within one week he died. Bill and I had a great time working together on the F-14 and I am thrilled he was able to know the chips worked the first time and that the F-14 had a very successful flight test. Bill has been missed greatly all these years and I had many thoughts of what we could have done together with all this new technology.

Around 1984 I had a computer retail store in Santa Clara, CA and a man walked in to buy some computer accessories. In discussions with him, I found out he was a F-14 pilot. This was the first pilot I had met and I was excited to talk to him and specially to ask about the Pilot Notification light. After explaining everything about the Pilot Notification light he informed me that the light was removed because *"no Navy pilot would ever abort a mission because a light came on."*

In the summer of 2016, I was invited to speak to the F-14 Pilot Association in Pensacola, FL. About 50 pilots and maintenance crews attended. During my talk, I mentioned this light and I was informed that it was changed to an amber color. I was also soon informed that a reset button had be added to my computer. At first, when I heard that a reset button was added, I was slightly insulted and did not know what to say. Why would they need to reset in-flight and what kinds of issues would that cause? After some quick thinking I asked, "Why did you have to add a reset button?" The answer was "The overall power provided by the F-14 became erratic and electrically noisy with added equipment and caused the Pilot Notification light to come on incorrectly."

The solution was to not believe the light until Reset had been pressed and then if the amber light stayed on to make decisions as if it was a real failure and not a falsely notified failure. That was probably the most brilliant solution to a very difficult situation. After my talk, I sat down next to a pilot and I asked him "Do you mean to say that while you are dog fighting you are pressing the Reset button?" He said, "Yes, all the time." I have been thinking of that scenario all the time since. Even though it would work perfectly it was a very clever fix for the F-14 power supply noise.

At the end, there were 712 F-14s built of which 435 were F-14A's. 102 F-14's was sold to Iran and today 20 of them are still operational. This fact contributed to the difficulty in the denouncing and releasing of my documents. I learned at the above talk to the F-14 Pilot Association that all F-14's used the same CADC and that no CADC was a factor in any F-14 crashes. Given the age of the technology those facts represent the excellence in the CADC design, the technology, and in the AMI computer chips.

Ray at the Pensacola, FL Naval Air Museum in front of F-14 #5. He personally tested this CADC before it was shipped to Grumman for installation. Of course, he is symbolically showing his baseball past. Photo Credit: Shauna Gregg (Photos with F-14) 2017

Chapter 4
Technology vs Forty Square Inches

In the following few pages, I list the most used technical names about my claim and explain in short what is the job of each of them.

A transistor is the smallest "smart" element on an integrated circuit. A microprocessor is made on one or more integrated circuits. The "smart" elements or transistors are combined to form logic gates which are the basic design elements in a computer.

A silicon chip and an integrated circuit are the same thing. A microprocessor is made of one or more integrated circuits.

The evolution of machine technology follows a simple path: mechanical parts, electrical controls, electronic control through large vacuum tubes, electronic control through solid-state transistors (100 times smaller than vacuum tubes), micro-electronic through chips (memory chips, micro processing units, 1,000 times smaller than solid-state) and today's nano-electronic (1,000 times smaller than microelectronics).

In its early stages, micro-electronics developed deeper and deeper levels of integration, passing from a small-scale integration, or SSI, to MSI (medium), and then LSI (the case for the CADC's chips), later VLSI (very large). After this way to define integration, the industry started giving directly the linear dimension of the basic part of a single transistor: first given in millionth of a meter, or microns, today (2022), many-billion-transistor chips use transistor size of 7 nanometers, or billionth of a meter. That is small!

The microprocessor executes instructions coded in the software memory. Simple microprocessors are called microcontrollers. Microprocessors, memory units, switching circuits, and other units are different kinds of chips.

Custom chips are ones made to a particular specification. For example, one calculator company would have custom chips made just for them and they could not be sold to anyone else. Standard chips were usually developed by the chip manufacturer (like AMI, Intel, National Semiconductor) and marketed to the general marketplace.

Many ways to build a computer existed. In considering the technology to use we looked at the currently available possibilities, TTL, MOS or P-channel MOS. TTL bipolar stands for transistors-to-transistors logic bipolar; MOS stands for Metal-Oxide Semiconductor. We did not consider the fastest TTL because it consumed very high power. The other possible technology was MOS logic: it was basically the same technology we used, but these were standard design modules that were already proven and developed. The reason we did not use them was because they were not dense enough in terms of the function that they provided: we only had 40 square inches to work with, so it would take way too many of these modules to produce the function that we needed. Ultimately, we were forced into using the very latest version of large-scale integration that was called P-Channel MOS. It only had about one year of field experience but it had proved itself well in high volume calculator chips. In reality, this was the only technology that could be used to design the complex computer that was needed for the F-14 content. We were approximately six months into the project when this decision was made.

Chapter 5

The Danger Zones

This is a more technical chapter than the previous ones. I wrote it thinking that everybody could understand it, but I am a tech guy and I could be wrong, so I apologize for any obscure passage you may find in the following pages.

"Danger Zone" was a featured theme in the movie "Top Gun" and looking back on the design and development of the F-14 microprocessor I can see several areas that proved challenging and could have stopped the development of the F-14 CADC and thus delayed the F-14 project or stopped it altogether. I thought it would be interesting to look back at these moments.

Ray Holt as a Young Engineer

The first "danger zone" moment is Me. Here I was first year out of college, three years from studying and almost graduating in Forestry, and with one computer class under my belt. I walked into a company thinking I was going to do one type of design and they moved me to a highly classified state of the art design that I clearly did not have experience in. Fortunately for me and the project, I was a highly detailed person and learned to love the process of digital design and the requirements to complete a properly working computer. Combining mathematics, engineering, and computers to a real-world solution was intriguing. The fact that computers either work or not was motivating to me because I knew eventually with enough effort that I could prove that it worked 100%. I accomplished this mission with the direct support of many team members, in particular, Steve Geller, Lynn Hawkins, Bill

Holt, Jim Lallas, and Murray Lubliner, and many other support people. The most amazing people to work with were the entire engineering staff at American MicroSystems who engineered the layout of the chips and manufactured them. The first chip set worked 100% but the program had one binary bit in a numeric constant out of 73,000: still not too bad for manual programming of all 73,000 bits! The programming took about two months. This one-bit error did not affect the early testing or delivery of the CADC for the F-14.

Project Schedule

The second "danger zone" moment was the project time frame. From the proposal delivery to project completion was 24 months. As a young engineer, I had no idea if that was possible or not, however, many experienced engineers and administrators had come up with that time frame. Looking back, I can imagine that they were quite nervous with the schedule and especially with the penalties for being late. I am happy I was not aware of that kind of detail. With the whole concept of a microchip computer being new, NO one had any real idea if it could be made or not. The assumption by everyone was it could be made and so everyone proceeded as if there was no risk at all. Any glitch in the design and/or programming and the manufacturing of the chips would delay the CADC 3-6 months. Fortunately, everything worked and delivery was on time. This is probably the most amazing part of the whole story. I am still amazed that no one put any pressure on me about the importance of the computer schedule.

Microchip Suppliers

The third "danger zone" moment was concerning our vendors. After we had a completed and proven design on paper, we had to get quotations from three vendors. The only three that could make the microchips were American Micro Systems, Rockwell, and General Instruments. The bid request included all six different chip types we needed for the project. The danger moment of this story is the fact that all three of the vendors turned us down! None of them wanted to make the chips. Now we were in a position of having a well-tested set of specifications for the microprocessor chips and no one wanted to make them, not even AMI who helped us design them. The main reason for the rejections was the fact that all three semiconductor companies were very busy with high volume manufacturing (like 10,000 – 30,000 chips per month) and the F-14 microprocessor chips were only going to have 100-300 made a month. This was a very small volume and no one was interested. The project was put on internal hold until this problem could be worked out. The CADC management asks me which vendor I preferred and, of course, I said AMI since they helped me with the chip design. Nothing was said for a week and then I was told to fly up to Santa Clara and start working with AMI on the chips. It was not until near the end of the project that I heard that the management of Garrett AiResearch negotiated a stock buyout of AMI in trade for a large amount of cash as well as AMI helping Garrett AiResearch build a semiconductor facility north of San Diego to be used as an independent second source supplier. This facility was called Garrett Microsystems. The Garrett Microsystems facility was later sold to Motorola.

Mean Time Between Failure (MTBF)

As mentioned above in my story, the next "danger zone" moment came when the Navy and Grumman asked us to provide the MTBF report on the CADC. MTBF stands for Mean Time Between Failure. This is a historical report on the failure rate of the parts in the CADC. With most of the CADC being new technology, especially the microchips, there was no history to provide. When this was passed on to the Navy and Grumman the project was literally put on hold for about two weeks. We kept working but as far as Grumman was concerned it was on hold. Not being able to provide an MTBF report was very serious. The military rarely, if at all, used technology that had no history. They wanted to know exactly the reliability of the technology.

With the microchip technology being barely two years old, no one had published any history on chip failure or reliability. Certainly, it was more reliable than any previous technology but there needed to be proof. During the two weeks on hold, there were lots of discussions between the CADC administration and managers and Grumman. I was only consulted near the end of the two weeks and I was asked if it were possible to "self-test" the microchips in flight. At first, I did not know what this really meant but after some of us discussed it I decided that it was possible to try. I did some research and I could not find how this type of self-testing was ever done. Eventually, I decided that I could program a series of computations and instructions that tested all the components or devices on the microchips. But I already wrote about this above and will repeat some of this important story.

For me, this turned out to be one of the most fun portions of the design. Basically, I had to determine mathematical calculations that would exercise all the transistors and all the connections on every microchip and to make sure that if any transistor or connection failed, either open or short, that the result of the calculation would be incorrect thus the microchip had a failure. This presented two very serious challenges; 1) How do I determine all the failure modes on these complex chips? and 2) what do we do when a chip does fail? The failure of the CADC was serious and would disable the operating of the F-14. To solve the first challenge, I asked for a programmer that would write a detailed microchip simulation for every chip with the program being able to simulate failures of open and short wires on every path on the chip.

The company decided to transfer my brother, Bill, from another department and have him work for me doing this simulation. It was wonderful working with Bill and he did a great job with the simulation. His work not only contributed to a very reliable CADC but also to a means that could be used to test these complex chips before delivering the CADC. The final run of his simulation took about three days on the AiResearch business computer and the weekend we needed to make the programming run. However, the company payroll was also due. We won and payroll was delayed until Tuesday.

Grumman and the Navy insisted that we 100% detect any single failure on any of the chips. Ultimately, we were able to do that for all but the 20-bit parallel multiplier and 20-bit parallel divider chips. These were very complex chips and there was not enough time to perform all the calculations to detect all the possible failures. Bill and I came up with a set of test calculations that would test 98% of the failures and we showed Grumman and the Navy that the 2% we could not test would only affect conditions that the F-14 was not designed to perform. Grumman and the Navy accepted our in-flight test calculations in place of the MTBF report.

However, the second question was still a problem. What were we to do if a failure was detected? A failed CADC would mean the F-14 would possibly crash. It was decided to put two microprocessor chip sets in the CADC, along with two sets of quartz sensors and if the first microprocessor was determined bad then it would be switched off and the second one would switch on thus allowing the F-14 to seamlessly continue its mission. The Navy also decided to add a light on the pilot dashboard that turned red when the 2nd microprocessor was enabled. This was called the Pilot Notification light. The pilot manual would train the pilot to then decide to continue the mission or to return home.

Prototype & Design

In order to make sure the microchips worked as designed and to make sure the design worked according to the specifications, a mock-up or prototype was made. This was an actual working CADC made from existing, standard, off-the-shelf electronic parts. Of course, it was not small but size was not our concern. The big concern was that the design and programming worked according to the specifications.

I don't remember the exact time but I do remember the prototype was basically working. The Admiral of the Navy in charge of the F-14 and CADC was invited to witness a demonstration of our working prototype. We were given about a two weeks' notice of this event. We wanted everything to look good so each day Lynn Hawkins and I would clean up the prototype which meant to make the wires look nice and for all the electronics boards to look good. One night I stayed late working on a small problem. It was about 10 pm and I was getting tired and was ready to go home, however, I wanted to make one more change to the design. In the process of making the change, I plugged the power supply in backwards (red to black and black to red) which in effect reversed the voltages to all the electronics. It all happened so fast I did not realize what I had done or the major damage I had caused. Soon I realized that I had destroyed all of the microchips or -as we would say- I burned out the whole prototype unit. I went home and did not sleep well.

The next morning, I told the manager of the CADC project, Phil Erath. Phil kept calm and said to go downstairs to purchasing and to have them purchase every available part in the United States that we used on the prototype and to have them shipped as fast as possible. Fortunately, every part was available and as they came in, we had "solder girls" or women that specialized in soldering on these circuit boards, remove and replace the parts. Eventually, we were able to replace all the parts and then we had to retest the whole prototype. Lynn Hawkins worked hard to make sure it all worked and as Lynn tells the story "When the Admiral was 50 feet from us the prototype worked." I am quite sure that Grumman and the Navy were not told about this and reading it here might be the first time they know. It was a very scary and nervous time for me and it might have cost me my job and Garrett some heavy penalties had the CADC not been demonstrated.

Programming the First Microchips

Once the microprocessor hardware microchips were defined, I had to consider the programming tasks. I knew ahead of time that I had about 600 calculations to perform. The CADC performed all calculations every $1/18^{th}$ of a second. However, I was to program it: I knew the task was huge and would take months. Most computers have what is called Software Tools for assisting in programming. The raw or lowest level programming is called binary. This is the native programming of any computer. Rarely does anyone program at this level because of the technical knowledge required of the hardware microchips. The next level of programming is called assembly language. At this level, the programming would write in short English words that are translated into binary. The highest level of programming is when the programmer writes in somewhat understandable English words and a specialized software called a compiler converts this program into the binary. Binary is the most efficient for programming but the most difficult and time-consuming. Using a compiler is the quickest way to program but also the most inefficient when producing the binary code.

The big problem with the CADC is that we did not have a compiler and would have had to create it before we even started programming. With only about three months to program, the decision was made to not write a compiler and to manually program in binary. These were very detailed and meticulous tasks and needed to be done in very quiet surroundings and not in the "bullpen" environment of our department. Most of the programming I did in the evening and during all night shifts. I did not mind working these hours as I needed the quiet hours.

Programming consisted of starting with the mathematical equations, determining the correct order of performing the calculations, translating this order into the real-time execution of the computer, and then actually writing the binary programming code (1's and 0's) in order to instruct the microchips to perform correctly. The entire program consisted of 62,092 1's and 0's. After I completed the programming, we had to make sure that all the binary code was correct. James Lallas wrote a simulator that would take the 1's and

0's and perform the functions of the computer. By using the simulator, we could make sure that all the mathematical functions were performing correctly using the design of the microchips. We could have spent months testing all the possible scenarios of the F-14 but we just did not have the time. In order to speed up the verification of the binary program, we did sample testing which involved testing all main function at their extreme limits, at several normal conditions, and in several random conditions. After about three weeks of simulation testing, we were convinced that the binary programming code was accurate. Because of the huge manual effort in programming, the possibility of being wrong or having problems was huge and the risk was great that some problem would prevent the first CADC from being delivered.

After I decided that the binary was correct, I had to translate the binary into the correct patterns for the Read-Only Memory or ROM microchips which stored the program. This process took about a week. After all the testing was completed, the final program was delivered to American Microsystem via a punched paper tape: all this was before flash drives, floppy drives, portable hard drives. Paper tape and punched cards were the only affordable way to transfer programs.

Now that the ROM bit patterns were delivered, it was about a three-month wait for the first set of chips. This was a very long time and all I could think of was "what if one bit was wrong and we could not deliver the CADC on time." Certainly, a very critical time. It was a go or not go situation. If the program was incorrect, it could be four months before we were able to get new ROM's. The good news is that only 1 bit was wrong out of 62,092 and that 1 bit was a data value for a math calculation and not critical to the early testing of the F-14. The logic design of the microprocessor chips worked perfectly. The CADC was delivered on time and worked beautifully. This was such a huge relief that I can still feel it.

After the first CADC was delivered it felt like smooth sailing for the next several months. Microchips seemed to work as expected and the big challenge was to get them into a CADC and to get the CADC tested and delivered. As I can remember, I only personally was involved with less than the first ten units. Garrett AiResearch had a very good testing department and I worked mostly with Ian Linton who picked up the workings of the CADC microchips very quickly. He was the only other person that could program the microchips and I heard he made a few revisions to the programming based on some early flight tests. Ian was a huge asset to the transitioning of the CADC from engineering to production.

F-14 Down

It was in December of 1970 when I was watching the news with my father and the video of the first F-14 crash was shown. This was the 2nd test flight of the F-14. As the pilots were making their approach for landing the F-14 lost control of the movable surfaces. This is what would happen if the CADC had failed. The plane quickly started losing altitude. Only a few hundred yards from the landing strip the pilot and RIO (Radar Intercept Officer) ejected. Fortunately, both were able to survive the low altitude ejection. The F-14 crashed and burned. My first thoughts were the CADC failed... new technology... new flight conditions... new everything. Why didn't the dual redundancy work? Could both computers have failed at the same time? Lots of questions and lots of deep thinking. I don't think I have ever felt such shock and nervousness as I did those next few days. I keep thinking it must be the CADC since so many of the CADC functions were involved in the movable surfaces. The next two days at work were very very somber and quiet. It didn't take too many days before Grumman announced that the central hydraulic hub had leaked. The CADC did not fail. Even though it was a huge relief to me I had a very sympathetic feeling for the designers of the hydraulic hub.

Chapter 6

My Claims

I would now like to talk about what I have called the General and Specific Design Accomplishments of the CADC microprocessor chip set. Any credit due any of these accomplishments is to go to the entire team that worked on the CADC including the engineering department at Garrett AiResearch and the engineering department at American Microsystems.

During the first few months of the CADC project (June 1968 – Aug 1968) I did extensive research on the state of the technology at that time. In my paper "LSI Technology State of the Art in 1968" which is available for download on my website http://FirstMicroprocessor.com. I documented most of what I was able to discover at that time.

One thing is clear from this research is that in 1967 high-density computer chips did not exist however they were expected to come soon. I suspect even while we were designing the microchips people were writing that such high-density microchips were not possible. Even in 1998 when the microchips were finally introduced, I think many did not believe it. Fortunately for those unbelievers, they did not profess their thoughts publically as my possession of the first set of microchips and the fact that the F-14A history is well known makes any denial a moot point.

There are lots of ideas and concepts that could be discussed on how the microchips evolved to their final configuration. Lots of design and environmental trade-offs were made and fortunately it became possible that high density microchips could be made at that time.

My <u>General</u> Design Accomplishment Claims Are:

1. *1st microprocessor chip set.* Not disputed with all the facts available.

2. *1st aerospace microprocessor.* Well documented.

3. *1st fly-by-wire flight computer.* Initially true. I understand later the pilot had the ability to override critical functions.

4. *1st military microprocessor.* Not disputed with all the facts available.

5. *1st production microprocessor.*

6. *1st fully integrated chip set microprocessor.* The six custom chip types contained all the control and computational ability necessary for the microprocessor to operate. This fully integrated capability was not accomplished in the commercial world until 1993 with the Intel Pentium microprocessor. The Intel 4004 required 53 additional electronic circuits surrounding it to operate.

7. *1st 20-bit microprocessor.* This is a claim that would have been true for about 20 years. A bit represents the accuracy of the arithmetic operations (add, subtract, multiply, divide). 4-bit means an accuracy of 16, 8-bits means an accuracy of 1024. 20-bits mean an accuracy of 1,048,576. The Parallel Multiplier and Parallel Divider chips performed 20-bit by 20-bit operations in 20 clock cycles. This was a marvelous feat even for non-LSI chips. Further technical investigations would show that the ability to perform at this speed required some rather

advanced techniques at the arithmetic carry-look-ahead level as well as with the layout of each arithmetic unit.

My Specific Design Accomplishment Claims Are:

1. *1st microprocessor with a built-in programmed self-test and redundancy.* As I discussed in my story the Navy and Grumman required that the microchips be self-tested in real-time during flight. We were able to accomplish 100% on all chips except the complex Parallel Multiplier and Parallel Divider where we accomplished 98%. Also, when a failure was detected, we switched to a whole new microprocessor, on a different circuit board. Basically, this setup was a self-fixing computer in real-time. I am not sure even today (2014) this has been accomplished.

2. *1st microprocessor in a digital signal processor (DSP) application.* A DSP is a particular application that requires lots of mathematical calculations. The first true single-chip DSP did not come out until ten years later with a hardware multiplier on-board.

3. *1st microprocessor with execution pipeline.* Execution, or instruction, pipelining is a design technique used to increase the speed of executing programming instructions. Instead of each instruction being executed one at a time, instructions are either broken in smaller commands or overlapped with the completion of the previous execution. The concept is to use the hardware as efficient as possible by speeding up the executing of programming instructions. Pipelining is rare in real-time computers. After the 70's pipelining was used in

most large-scale Super Computers but rarely, if at all, in microprocessors.

4. *1st microprocessor with parallel processing.* Parallel processing is somewhat similar with co-processing. The concept is that a large, or time-consuming, mathematical equation is sent to a different set of hardware for computation while the main CPU continues with other tasks. The CADC Parallel Multiplier and Parallel Divider did just that. Both of them were 20-bit true parallel computational hardware microchips. The concept of chip co-processors was new and ten years before Intel introduced their 8-bit co-processor, the 8087.

5. *1st Read-Only Memory (ROM) with a built-in counter.* A read-only memory contains pre-stored data and cannot be changed. This data can be programming instructions or fixed data used in mathematical equations. The CADC used ROMs in both capacities. Each piece of data in the ROM must be addressed in order to access the information. Typically, this addressing capability is implemented in the CPU function, however, doing it this way causes multiple wires or connections from the CPU to the ROM. In order to save these connections, we put the program counter inside the ROM and then we could increment it or preset it. This was a very unique implementation that really helped the microprocessor chip set to be as small as possible.

With the CADC out of engineering and into production, my work slowed down dramatically. Some days were just boring. Some members of the CADC team were staring to leave for other jobs. Tom Redfern went to AMI, Russ

Almond, Cameron Pedigo left to sell semiconductors. I was still a new and somewhat inexperienced engineer and I did not know much about changing jobs so quickly. I was thinking that unless I was fired that I would have a job for a very long time. I was asked by the CADC project manager, Phil Erath, if I could stay until the follow-up documentation was completed. Of course, I would stay as I didn't even know I should be considering leaving.

What Next?

The next year (1970-71) was spent cleaning up all the documentation of the F-14 CADC microprocessor. This involved the microchip design as well as the programming. I ended up writing the Technical Manual for the CADC microchips and documenting how the microprocessor worked and how the programming worked. Even though writing every day was not as exciting as creating an engineering design, I did find it satisfying that everything was in one document. I also gathered some of my design notes and put them in a binder. I was able to keep, as my personal collection, the first microchips off the production line, a copy of the Technical Manual and my Engineering notes. These I still have with me today and I show them to high school and university students when I speak. I also have an F-14 pressure sensor that I purchased several years ago and I also display it during my talks.

During the year of documenting, I wrote a technical paper on the CADC microprocessor. It was my very first technical paper. At first it was difficult to figure out how to present the CADC details. I spent time reading other research and technical papers and eventually became comfortable writing. When I thought the paper was near completion, I submitted an early copy to the Computer Design magazine. This was the premier magazine for computer designers and if I could get an article accepted it would be a huge advantage to my career. I did get the paper accepted but they told me that because it was obviously a military project that I had to get approval before they published it. I did not think it would be too difficult to get approval.

I presented the paper to the CADC project manager, Phil Erath, and he immediately said NO and preceded to explain to me the importance of the confidentiality of the workings of the CADC. I politely asked him if we could try and see if Grumman or the Navy would approve the paper and he agreed. In a short period of time, I think within a week, we received a NO from Grumman and the Navy. I was extremely disappointed that I could not talk about it. During this time, I had also asked Garrett AiResearch if we could patent the microchips and we again asked Grumman and another NO was given but this time with an explanation that all patents were public domain and that the details of the CADC were not to be made public. At the time I did not know the full impact of these decisions but within four years it hit me as I started to see semiconductor companies introduce their own versions of a microprocessor.

1998, the end of an oblivion

In the last many years that I have been able to talk about the CADC microprocessor, I get asked a lot if I knew the huge impact this design would have on our society and on engineering design. On the other hand, I also get asked what difference this design made since it was kept secret for 30 years. I will attempt to answer both of these questions.

My first public appearance as the designer of the world's first microprocessor happened during the Sept. 26, 1998, Vintage Computer Festival event. After I got approval from the Navy, I saw a flyer about this festival and I contacted the organizer and we had lunch. He listened to me but was not sure to believe me or not. I told him I had actual chips and documentation and we met again the next day and he was shocked and became a believer. He said he wanted me to be his main speaker at the upcoming Festival and he arranged the Wall Street Journal ad and my 5-min spot on NRP radio. There were about 75 people at the talk: among them my family, James Lallas and Tom Redfern from the design team, a few of my other friends and lots of people I did not know. It was about an hour-long PowerPoint presentation and I showed my artifacts. It was really nice to be able to talk about it and lots of details came back to me 30 years later. I remember the Wall Street Journal writer challenging me on some points and Tom Redfern said that "if Ray says it is true then it is true." I think the quote was written up in one of the articles.

All the details of the microprocessor have been available on my website for over 20 years and still writers will say in articles that the microprocessor is unknown because of its 30-year secrecy. As of this writing the detail workings of the microprocessor are being software simulated and I have decided to reproduce the microprocessor in standard off-the-shelf digital circuits.

Ray Holt's Photobook

First row: The F-14 CADC in its final package; the F-14 CADC Prototype (1968/1970).

Second row: Ray holding one of his chips in front of an F-14 Tomcat; Ray with his pupils (2010); the MP-944 Original Technical Manual.

Third row: Bill Holt; Ray holding one of his chips close to an F-14 Tomcat; Ray's Press Card as a journalist; AMI 7200 Microprocessor Prototype.

Fourth row: Ray holding one of his chips near an F-14 Tomcat (see page 126); Robart I - 1st military robot using SYM-1; Jolt series add-on boards, Ram, and Audio.

Photo Credit: Shauna Gregg (Photos with F-14)

©2006-22 Copyright by Ray Holt All Rights Reserved   2nd Edition
This publication cannot be reproduced in any form, without the written permission of Ray Holt.

Chapter 7
Beyond the F-14: More Microprocessors

In 1971 I was asked by Mr. Ken Rose, the Director of Engineering at AMI, if I would consider working for AMI in Santa Clara. The timing was good. My documentation was completed and there was no new project to work on. Because of the legal relationship between Garrett AiResearch and AMI, I made a formal request to be allowed to enter into discussions with AMI. My request went all the way up to the President of Garrett, Mr. Harry Wetzel. My request was approved and after a few trips to AMI, I decided to accept their offer as a System Designer for Calculators. I would work for/with Brian Shubert and James Kawakami. Both were excellent system chip designers and most of their chips worked the first time. However, designing calculator chips was a little difference than general purpose computers. It wasn't quite like joining Garrett for the first time, but very similar. A new style of working environment, engineering work I had not done before, and microchips you could not see. This time I was on the reverse end of the Garrett – AMI relationship.

Calculator chip design was different than the computers in that the calculators do exactly the same thing for the different math key strokes. It just does it on a different set of numbers. The control logic for a calculator is called a State Machine, a style of computer design where one stage of operation depends on where the previous style ended up. The control sequence was predetermined and would work the same every time. James and Brian always seemed to come up with unique design techniques that saved chip real estate and increased speed. Brian seemed to become a specialist in the keyboard scanning logic and had one particular chip that was always used for that purpose.

During my second year, AMI hired Gaymond Shultz. Gaymond was a computer designer hired to work with me to come up with a microprocessor for AMI that could compete with the Intel 8080. Our design was called the AMI 7200 for 1972. It was a very general purchase microprocessor even to the point that the micro instructions could be changed. In 1973 I wrote a paper on the AMI 7200 for the IEEE Western Region Conference on Minicomputers. The paper was called "Architecture and Organization of the AMI 7200: A New Breed of Minicomputer." My paper was accepted for the conference in Honolulu, Hawaii. I took a second week as vacation. My first time in Hawaii. It was really nice ... and expensive.

In 1973 we worked on the AMI 7300, a much more ambitious design as this microprocessor was to contain not only the arithmetic logic but RAM and ROM all on the same microchip and it was to be 100% microprogrammed microcode. In other words, the main instructions could be completely changed by uploading microcode that redefined the main instructions. This made the microprocessor a super power microchip and was versatile enough to satisfy just about every project need. Another engineer, Harold McFarland, was hired to work on the microcode. He and Gaymond were to work together, which revealed to be a huge mistake. Gaymond was an experienced hard-nosed west coast engineer, and Harold was an inexperienced (as far as microchips) hard-nosed east coast engineer. I usually didn't stay around when they were resolving an issue. Somehow with all the personalities involved, we were able to complete the design and to get it into production.

Soon after this, the infamous decision by AMI Marketing that "there is no future in microprocessors" was made, and our 25-person microprocessor group was disbanded. Gaymond and Harold left AMI fairly soon. I left within a few months and became a business partner with Manny Lemas who was hired to help us finish the AMI 7300. Manny and I worked for a consulting company called Compata Corp. Through Compata we were able to get a contract with Intel to teach some local courses on the 4004 and 8008. Later that year we contracted with Intel for the two-year training sessions around the country.

At the same time, AMD was thinking about starting a microprocessor group and hired Manny and me for a day to discuss the dos and don'ts and the timing of such a startup. We spent most of the day with Founder and President Jerry Sanders and his marketing team. It was decided by AMD that the time was not quite right for them as they were heavily committed to their Mil-Spec line of chips. Within two years AMD did start a microprocessor division and hired James Kawakami and Brian Shubert and AMD became a very formidable competitor to Intel. Brian later went to Intel to run their Graphics Division.

Chapter 8
Microprocessor / Microcomputer Projects
The Trash-80

Sometime around 1975-76, Manny and I were asked to develop a prototype desktop computer for Radio Shack. We built the prototype around the Intel 8080 and packaged it with a keyboard and monitor built into one case. This was a fairly new approach to desktop computing. We even had the computer start-up to the Microsoft Basic language. Radio Shack loved our design and asked if they could keep our prototype for a week. We got it back about six months later. Most wires had been displaced and moved which would have been the result of trying to reverse engineer the design. Within six more months, Radio Shack announced the TRS 80 Model I. We thought it looked and operated very similar to our design and prototype, and the subtle proof of that was the unreliability of the computer. Reverse engineering a prototype is not a good idea. For this reason, the TRS 80 Model I received a well-deserved nickname of TRASH 80.

"On the Edge – The spectacular rise and fall of Commodore", by Brian Bagnall (Variant Press, 2006) is one of the best books ever written about that time period. Bagnall dedicates a chapter to investigate the true story of the TRS-80 home computer. The chapter holds the name "The Trash-80", using the nickname given at that time to that device. Let's read it from the beginning.

> "The undisputed leader of the first several years was the TRS-80 from Texas based Tandy Corporation, owners of the Radio Shack chain of stores. John Roach, the man who viewed Chuck Peddle's

> PET 2001 prototype at the January 1977 CES, led Radio Shack into computers.
>
> According to the official Radio Shack history of the TRS-80, a Radio Shack employee named Don French, initially conceived the idea of selling computers through Radio Shack in 1976. By December 1976, the designers had the official go-ahead to develop a computer for Radio Shack."

This is the official version. But, as always happens, truth is not always the same for everybody.

> "This version of the story seems peculiar, since Roach gave Commodore a bid for the computer contract at the January 1977 CES. Perhaps John Roach had not intended to use the PET at all and just wanted to evaluate the competition. According to the history, engineers finished their handmade prototype by January 1977."

Bagnall digs more deeply into the mud. In the footnote, n.4, starting on the same page, he adds more relevant details.

> "At least two other companies received bids for the Radio Shack computer: Sphere Corporation and Microcomputer Associates. In later years, Michael D. Wise of Sphere claims Radio Shack lifted parts of his design. Ray Holt of Microcomputer Associates (...) also believes the prototype he offered was reverse engineered".

Chapter 9
JOLT and Super Jolt: Birth of a Home Computer

When discussing our second contract with Intel, we started Microcomputer Associates Inc. (MAI), our own company, that Manny and I owned until 1978. MAI started on a financial shoestring. I think we had $5,000 from the Intel contract. While we were able to get many small consulting jobs, we also decided to create two types of products.

The first of our two new products was the Microcomputer Digest. This was a 16/24-page digest of news and announcements of everything in the microprocessor and microcomputer industry. This was the first small computer publication. We charged $60 a year and after one year we had over 1,000 subscribers. We all enjoyed the Digest but after two years it was such a huge task to put out each month, we had to ask ourselves if we were engineers or publishers. We chose engineering and proceeded to try to sell the Digest. I know we talked to three major publications but I remember the reply for two of them. These were McGraw-Hill and McMillian. Both replied with similar comments, such as "there is no future in this market" and "your subscription base is too small". If only they could have known that within two years the small computer publication market was huge and almost outgrew them. Either one of them could have been the first in the market, starting with 1,000 subscribers. Just like AMI, both McGraw-Hill and McMillian made a huge mistake. Original copies of the complete two years of the Microcomputer Digest are still around with our Editor, Mr. Darrell Crow. All the issues of the Microcomputer Digest are located here:

©2006-22 Copyright by Ray Holt All Rights Reserved   2nd Edition
This publication cannot be reproduced in any form, without the written permission of Ray Holt.

https://archive.org/details/bitsavers_microcompuomputerDigestv01n06Dec74_2792467

But we were designers, so the most important project was a series of computer kits that the end users could assemble and have their own computer at home. We called the series The Jolt. I think we had 7 cards in the series: CPU, RAM, ROM, Power Supply, I/O, Audio, and the Universal card. We made a huge international advertising splash and eventually sold over 5,000 kits. This was the first real home computer.

Manny and I spent many hours trying to figure out a good name for this computer. Eventually, and I am sure out of frustration, Manny finally said "*let's name it Jolt*". Since Manny was part Mexican and spoke Spanish, he always called me Jolt, Spanish for Holt, so it was now the name of our computer

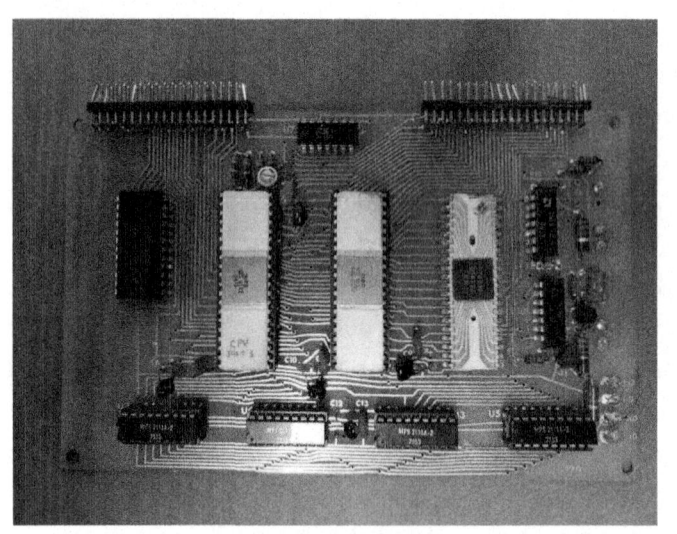

JOLT Computer Board #1

product. After the JOLT kit series, we sold the SUPER JOLT card as an OEM computer to be used inside equipment. We sold over 10,000 of these.

©2006-22 Copyright by Ray Holt All Rights Reserved   2nd Edition
This publication cannot be reproduced in any form, without the written permission of Ray Holt.

Besides the JOLT, SYM, and the Microcomputer Digest we received many contracts for microcomputer applications. Several that I remember are the Lucky Dice pinball game using the Intel 4004 (first microprocessor pin ball game which I programmed), a hand-held chess game, an earthquake detection and measuring device, a computer interface controller card for the first moon rover for Jet Propulsion Labs, a microprocessor-controlled microfiche machine, and several high-end calculators. All in all, I think we consulted or helped design over 50 new microcomputer-based projects.

Chapter 10

VIM-1, SYM-1, & SYM-2

Our product line continued with the VIM & SYM series. Chuck Peddle of MOS Technology and designer of the 6502 microprocessor was a good friend of my partner Manny Lemas. Chuck designed and introduced the successful KIM-1 development board for the 6502, however, it had several limitations. We asked Chuck if he had plans to expand the KIM-1 board and he said no. He highly approved of our designing and introducing the VIM-1 and SYM-1.

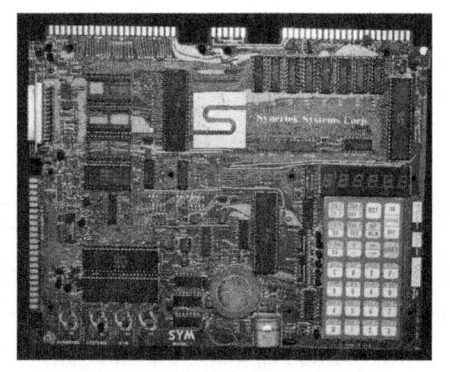

SYM-1 Computer Card

The VIM-1, Versatile Interface Module, was a very enhanced KIM-1 with added memory, input/out ports (connections), and an oscilloscope trigger connection. After Synertek bought Microcomputer Associates Inc and renamed it Synertek Systems the VIM-1 was renamed SYM-1. The SYM-1 had a look up table software vectoring feature. Wikipedia said about this feature on its SYM-1 webpage, "Some of the other computer designers of this era failed to grasp the significance of this elegant use of vectors to the software mapping of new developments in hardware."

The SYM-1 sold over 50,000 units and was very popular in teaching college engineering students how to program. The SYM-1 was a complete computer

©2006-22 Copyright by Ray Holt All Rights Reserved   2nd Edition
This publication cannot be reproduced in any form, without the written permission of Ray Holt.

on a single card and could easily be mounted inside an enclosure as the central processing unit. Two of the systems using the SYM-1 were the Robart I and Robart II robots designed by Bart Everett as a postgraduate student for the US Navy. These were the first two military robots with a microprocessor. Bart's story is mentioned in the second foreword of this book.

Our quick success with the JOLT and SYM cards turned out to be our challenge. With production of the cards increasing, we started running short of cash to purchase parts. At the same time, a two-year-old semiconductor company called Synertek (and supplier of chips to Apple) was looking for a group to make computer cards from their chips. We were already using their chips and we were available, so within a few months we were bought out by Synertek and became Synertek Systems. One month after Synertek bought us, Honeywell Information Systems bought Synertek so within one month we went from a small computer company to the smallest high-tech division of the multi-million- dollar Honeywell. Honeywell kept us a division separate from Synertek. This was in 1978. [Note: In 1985, Allied Signal purchased Garrett AiResearch and then in 1999 Allied Signal merged with Honeywell and kept the Honeywell name.]

With Honeywell money we were able to ramp up marketing, advertising, and production as well as hire a few more engineers and programmers. This growth was nice to see after struggling on limited funds for several years. The obvious disadvantage of having a company like Honeywell on top is the constant meetings and reports. Eventually, after two years I grew weary of all the non-engineering work and ventured out on my own. My new venture was called Cornerstone Computers (dba Cornerstone Software, Cornerstone

Business Services). This was April 1980. I am still operating under these names 40 years later.

After I left Synertek in 1980, Synertek Systems introduced the SYM-2. It offered a choice of three microprocessors, an on-board power supply, eight toggle switches, and eight LEDs both for enhancing the input/output capability. The SYM-1 and SYM-2 each sold for around $249. Extensive information on the SYM cards can be found here.

http://retro.hansotten.nl/6502-sbc/synertek-sym-ktm/sym-1/

Chapter 11

The Christian Athletic Association Inc (CAA)
... and other ventures

In 1977 I started a Christian non-profit sports league called the Christian Athletic Association, Inc. (CAA). CAA ran year-round sports leagues for youth 4th-12th and adults. Almost all the games were on Saturday with coaches' and parents' meetings during the week. By 1980 the organization grew to the point where it was almost full-time work on a volunteer basis. This was part of my decision to leave Honeywell as I needed to decide if I wanted to pursue long hours as a high-tech manager or work with youth and try to help them with a positive and encouraging sports experience. I continued with CAA until the year 2000 when it had about 1,000 kids participating year-round. My good friend and youth worker, Jonathan Detweiler, from Mississippi took over as Executive Director for the next five years. It is still going today (http://PlayCAA.org)

When I left Honeywell, my income dropped significantly. Even though I was very sure of my decision I really tried hard to make enough to keep the bills paid and not to have added pressure of living on the edge. I am not sure I ever reached the comfortable point as I was rarely able to reach half of my high-tech income.

Software Distribution, CP/M, and Gary Kildall

One of my first new ventures was distributing software. This was a brand-new type of business. Consumer and small business software was a growing market especially for the CP/M operating system developed and sold by Gary Kildall of Digital Research. I decided to distribute CP/M based software for all the many types of computer systems that could use it. I sold under the name Cornerstone Software. I was the 2nd software distributor next to a much larger software distributor called SoftSel. I was able to survive selling software for about two years until larger and better-financed companies entered the market.

After this most of my work consisted of computer and network installations for professional offices such as medical and dental offices. I opened a computer retail store in Santa Clara, CA to have greater exposure and space to work and to teach classes but the overhead costs were quite high and on a lot of days I spent time chit-chatting with customers that never bought. Computer retail was not all that great of a business. After ten years, I closed the retail store and continued my network installation business out of my house. I was able to be a local representative for some specialized software for video stores and this helped me to get some large contracts for video store equipment, software, installation, and warranty. By the late 90's the video business changed and most of my video store clients either went out of business or could not afford to hire me.

Graph-on, terminals... terminated

In the early 90's I worked for a good friend, Walt Keller, as Director of Information Systems and Customer Service. His company was called GraphOn Corp and they made computer terminal hardware and software for a variety of networked systems. The company did well but barely could survive in the up and down terminal market. After two years the company decided to get out of the terminal hardware equipment business and to concentrate on the software. At this point, I was not needed and I went back to consulting and network installations.

StartHere.com

Fortunately, in 1995 my oldest son Mark urged me to start using the Internet for email and some basic searching. This got me into Internet Marketing early on. After Yahoo, I was the 2nd one to have an extensive index search page called StartHere.com. I kept the name until 2012. I also started doing webpages and Internet Marketing and continue that today.

Chapter 12

What is Ray doing now in 2022?

In 2010 I ended up in Mississippi at the request of a friend, Dr. Dolphus Weary. Dr. Weary was President of R.E.A.L. Christian Foundation and wanted me to travel the rural roads of Mississippi and visit the ministries and afterschool programs he supports and repair computers. I did this for two weeks and was able to fix each of the problems. We had lunch after two weeks and he asked me "What do you think about doing this kind of work?". I hesitated a little while and then said "I think I like it." Meaning, I liked helping rural students that had low opportunity to learn about technology.

I committed to stay in Mississippi for one year to see if I could develop some kind of program for the rural kids and also continue helping the afterschool programs. I decided to start a robotics class in a town called Mendenhall and at the same time setup webpages for the afterschool programs. Eventually the robotics class was stopped as kids quit coming. This was due to no local support and no parental support. The kids had other options than attending a class on their own time.

The second year I met Pastor Tony and Shannon Duckworth from Mount Olive Ministries out of Mount Olive, MS, a small town of around 900. Tony and Shannon have a group of about 30-50 families that they work with. After a discussion with Tony, we decided to start an all-day Saturday robotics program for 5th – 12th grades. Tony was able to get a small grant for equipment and for some of my time. We continued the Saturday robotics program for three years. During this time the kids had success in placing state-wide in several robotics events. Our second year the kids took second in State and went on to World Competition and took 6th in their category. These successes encouraged the kids and many families. In 2014 Tony and Shannon started a private school and continued the robotics program through the school.

I decided to expand beyond the small rural towns and to start a statewide program. I formed a non-profit organization (NGO) called STEM Advancement Inc and started operating in Mississippi as Mississippi Robotics. Mississippi Robotics is committed to providing hands-on experiences and educational opportunities in math, engineering, and robotics to low-income, disadvantaged, students across rural Mississippi. During the school year each week I travel the rural roads of Mississippi and visit 3-4 rural schools or afterschool programs. Twice a year, in November and March, Mississippi Robotics runs a statewide competition challenging the kids in a variety of STEM events from math, science, robotics, and Mechatronics. Our first competition, in 2014, had 39 kids and our most recent competition drew over 300 kids from 30 organizations. We are now running five competitions annually.

Mississippi has over 400,000 students and over 200,000 live in poverty. These are the kids that Mississippi Robotics is trying to reach with technology driven programs that introduce and teach them "employable skills" in hopes that they can get jobs that pay well and can help them and their families get out of poverty. Eight of the 10 students that went to the World Competition in 2013 are now in college with six of them pursuing a STEM program of study. One of the girls graduated from Ole Miss in Electrical Engineering several years ago.

Chapter 13
Vision: STEM and Robotics in the 21st Century

Computers are machines that do two things well: perform calculations and perform decisions. The faster computers perform calculations and decisions faster. It's that simple. There is no limit for the speed of calculations and decisions from the computer designer viewpoint. A design that works just needs to go faster. Technology determines where we are at any given point in time. In 1968 from 375KHz to 2022 at 5GHz is an increase of over 10,000 times. Why do we need more speed?

Computer chip performance or capability if measured by the number of transistors has increased from 3000 transistors in 1968 to 3 billion transistors or more in 2022 an increase of 1 million in density. Why do we need high performance in small chips?

WHY are we wanting to go faster and have it smaller? In 1968 there was no vision or prediction that computers would penetrate our daily lives, especially in 50 years. Did we change our minds on what computers are for? Did civilization all of a sudden take control of human existence? Why do we need such computing and decision-making power? One generation… our parents survived without them and now we think we can't? What is this driving force?

The driving force is our unlimited need to have more and more data. Text, video, and color in all forms and all in an instant and at our fingertips. We have become gluttons of data. Data that takes billions and billions of transistors to collect, analyze, and report and then we throw away the results and ask for it again. Nationally we build huge buildings with multiple petabytes of storage just to save years and years of data just so we can analyze it in the future... phone call, texts, emails. Why? We are not satisfied with communicating at today's speeds, we want it faster. We want more and richer data in an instant, now.

The next generation will chuckle and laugh at all of this. What's the big deal? The world will literally be at their fingertips. Everything uttered, declared, built, and destroyed will be instantly available to everyone. What next?

Microprocessors. Machines that collect, analyze, and report data at super-fast speeds so humans can be satisfied with all knowledge in an instant. That is what we have created. When we reach the ultimate of this scenario, then humans will have become non-thinkers, and non-decision makers. Computers will have human's dependent on them. Our brains will not be able to comprehend what we have created in this machine that can calculate and make decisions at super normal speeds.

Fortunately, at this time, we are not there yet. We can still think for ourselves. The non-thinking generation is alive and well as they are our present K-12 students. These students are dependent on their smart phones and the Internet and all its social services. Library research, visiting people face-to-face to get to know them, writing letters is unheard of. Learning subjects at school... Why? It's all on the Internet. I just speak into my smartphone and it's all there.

The Internet is like one big microprocessor (actually millions of decentralized microprocessors) that control Things (thus Internet of Things - IoT) connected through unique addresses (IP addresses). From a dedicated microprocessor chip set in 1968 to millions of connected remote microprocessors, we have progressed to a global computer-controlled world. The "things" of IoT can be anywhere in the world and connected. It's an amazing thought to have each piece of your business, your security system, your warehouse network, your automobile company, etc., all connected as one big computer system streaming through unknown servers located in unknown places and all secure from others. We are very close to completely depending on a complex network of IoT's that are out of our control. Has the Internet made humans dependent on it?

The education of our future scientists and engineers is all of a sudden important as we have realized that a whole generation of youth have skipped the basics of science, technology, engineering, and math (S.T.E.M.). They became users and not creators of STEM. Now STEM is a hot buzz word in education and we are, almost desperately, trying to play catch-up with the current generator of youth. The problem is that they are not interested in learning how to create as they are enamored and stuck on using technology. Our youth are plugged into using technology... texting and video consumes 15-25-year-old people. They have no idea how all of this is created. A decreasing number of students are entering engineering and even less graduating. The USA is experiencing a huge void in filling hi-tech jobs, especially ones that involve any kind of programming. Many companies are finding it extremely difficult to fill hi-tech job openings. The answer in some industries is robotics.

Robotics is the field of computerized machines performing dedicated tasks with high performance and precision. Robots don't sleep or require time off or need benefits. The use of robots in industry makes sense due to the lack of skilled labor. Robotics is leading the STEM education field. Students like to build and program robotics. There is great satisfaction in creating a machine that does what you want it to do. The design and application of robots in all fields is growing.

In Mississippi, one of the lowest economic and educational states in the USA, I have been introducing STEM and robotics to a very large number of rural students. Students from very small communities with low funded schools to the few urban cities. The lack of equipment and skilled teachers prevent many of these students from learning STEM in their public schools. I have learned that none of these rural students lack the ability to learn, just the opportunity to be taught and the lack of equipment. In very selected rural areas of Mississippi 3rd to 12th grade students are excelling their knowledge in STEM and robotics and will become the designers and users of the new hi-tech skilled workforce. I am teaching "Employable skills" that apply directly to manufacturing and producing our products.

The microprocessor, our very hi-tech electronic manufacturing, the Internet, the IoT, the connected devices, have changed the way we live, communicate, and connect with each other. It's unfriendly and impersonal but it's efficient and fast and we seem to like it. We like being dependent on the computers of technology that do all this work in the background just to satisfy our need for instant knowledge.

Is the "computer" really unfriendly? I would like to challenge this statement: this landscape is quickly changing because of robots and smart machines. Human-computer interaction is growing exponentially and they are becoming more personable. The computer we are used to, is not the one the kids will be dealing with in the future.

In 50 more years, what will it look like and will we even recognize what we have now? I think not!!!

2020 ... Happy 50th Birthday to the Microprocessor age.

2020 ... Happy 50th to the F-14 Tomcat

Two gifts we did not see coming!!!

©2006-22 Copyright by Ray Holt All Rights Reserved   2nd Edition
This publication cannot be reproduced in any form, without the written permission of Ray Holt.

Ray Holt visited Italy in 2017. Leo Sorge arranged his stay, five lectures with various environments (IoT, start-ups, vintage computing, makers, and university), and his tour of Rome beauties and surroundings.

©2006-22 Copyright by Ray Holt All Rights Reserved   2nd Edition
This publication cannot be reproduced in any form, without the written permission of Ray Holt.

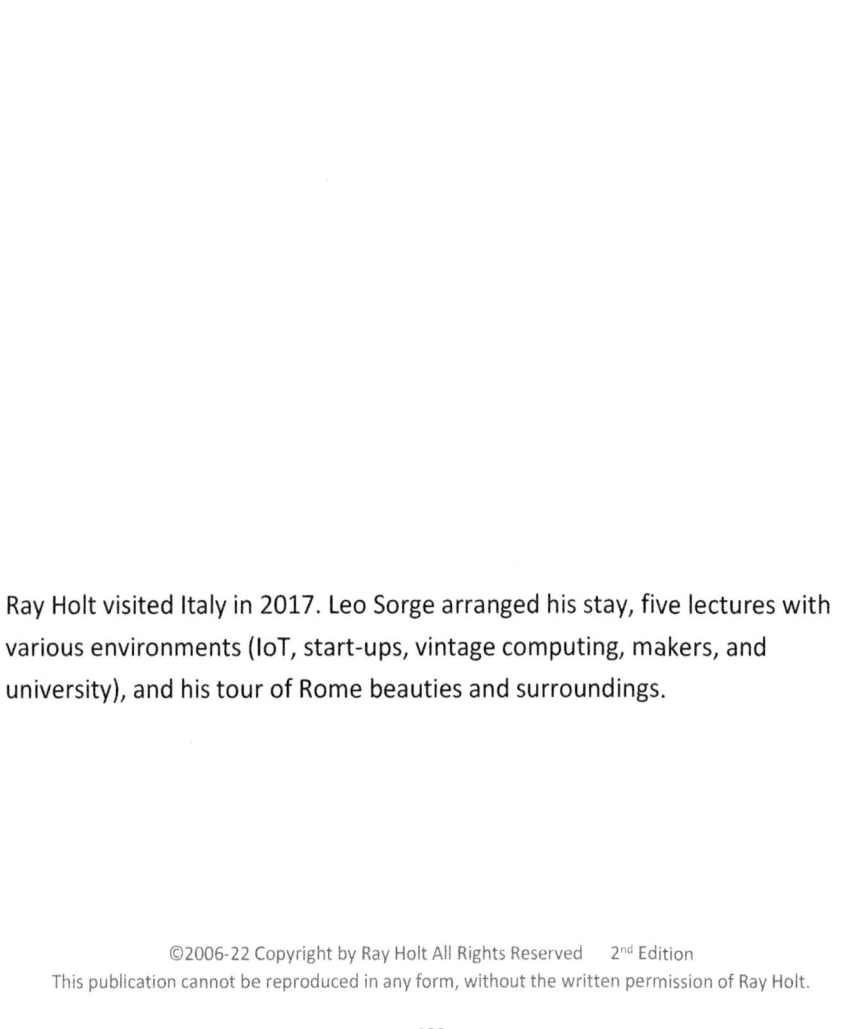

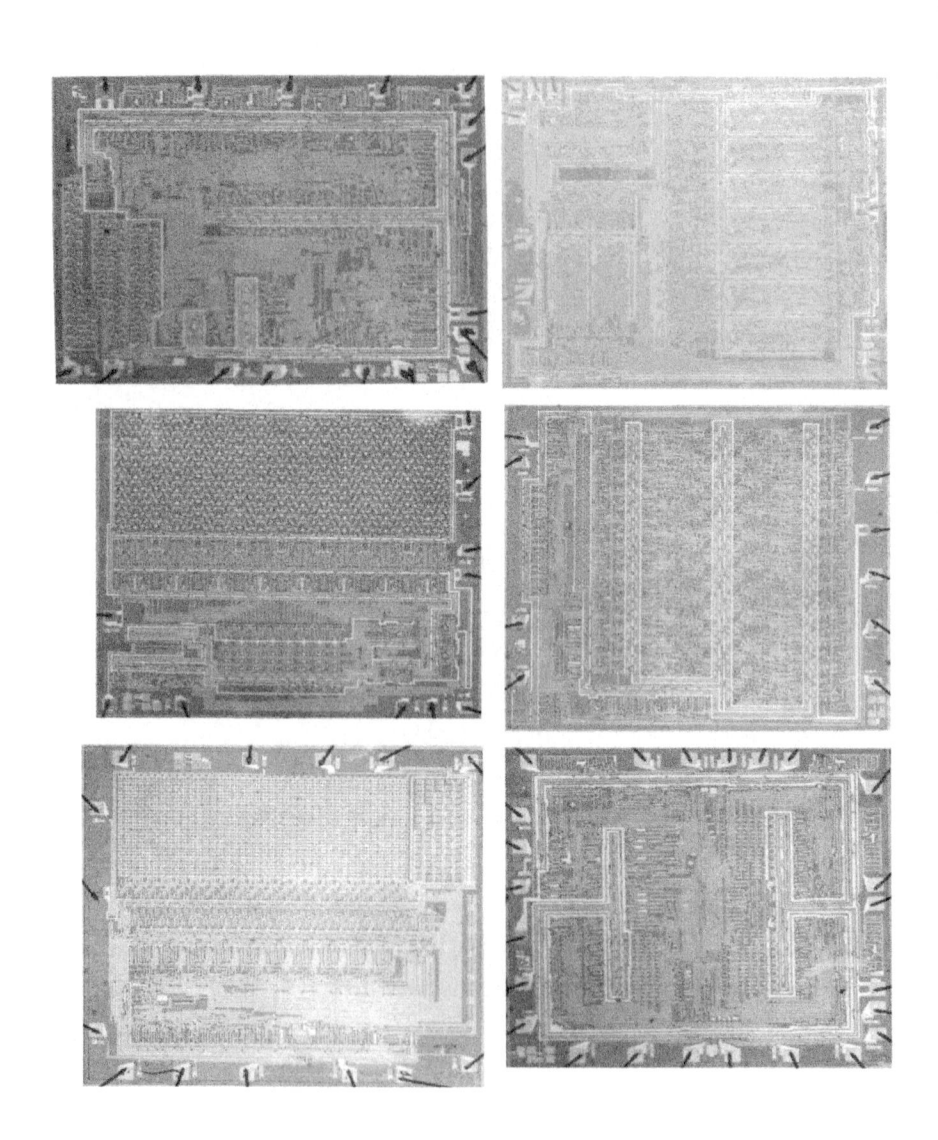

©2006-22 Copyright by Ray Holt All Rights Reserved   2nd Edition
This publication cannot be reproduced in any form, without the written permission of Ray Holt.

The full chipset:

F-14 CPU/Special Logic PN 944113, F-14 Parallel Divider Unit PN 944112,

F-14 Random Access Storage PN 944114, F-14 Parallel Multiplier Unit PN 944111,

F-14 Read Only Memory PN 944125, F-14 Steering Multiplexer PN 944118.

Ray Holt with F-14A at the Pensacola Naval Aviation Museum, Pensacola, FL
Photo Credit: Shauna Gregg (Photos with F-14)

©2006-22 Copyright by Ray Holt All Rights Reserved   2nd Edition
This publication cannot be reproduced in any form, without the written permission of Ray Holt.

Printed in Great Britain
by Amazon